T0295147

Microbes

Microbes

The Unseen Agents of Climate Change

DAVID L. KIRCHMAN

OXFORD
UNIVERSITY PRESS

Oxford University Press is a department of the University of Oxford. It furthers
the University's objective of excellence in research, scholarship, and education
by publishing worldwide. Oxford is a registered trade mark of Oxford University
Press in the UK and certain other countries.

Published in the United States of America by Oxford University Press
198 Madison Avenue, New York, NY 10016, United States of America.

Library of Congress Cataloging-in-Publication Data
Names: Kirchman, David L., author.
Title: Microbes : the unseen agents of climate change / David L. Kirchman.
Description: New York, NY : Oxford University Press, [2024] |
Includes bibliographical references and index.
Identifiers: LCCN 2023042495 (print) | LCCN 2023042496 (ebook) |
ISBN 9780197688564 (hardback) | ISBN 9780197688571 (epub)
Subjects: LCSH: Bacteria—Climatic factors. | Microorganisms—Climatic
factors. | Carbon dioxide mitigation.
Classification: LCC QR100 .K59 2024 (print) | LCC QR100 (ebook) |
DDC 579/.1727—dc23/eng/20231212
LC record available at https://lccn.loc.gov/2023042495
LC ebook record available at https://lccn.loc.gov/2023042496

DOI: 10.1093/oso/9780197688564.001.0001

Printed by Sheridan Books, Inc., United States of America

Contents

1. Introduction   1

2. The Natural Carbon Cycle and Fossil Fuels   12

3. Carbon Sinks and Sources on Land   29

4. Carbon Pumps in the Oceans   48

5. Clouds, CLAW, and a Greek Goddess   72

6. Slow Carbon and Deep Time   87

7. Natural Gas   108

8. Laughing Gas   130

9. Microbial Solutions   151

Acknowledgments   177
Notes   179
Glossary   215
Selected Bibliography   219
Index   225

1

Introduction

Separated by only an hour-and-a-half drive on the Big Island of Hawai'i, the climates of Hilo and Kailua-Kona couldn't be more different. Located on the east coast, Hilo is cooler than a tourist may expect for the Aloha State, and the days are often cloudy and damp regardless of the month. Many hours of rain sustain lush forests, diverse botanical gardens, and dense groves of macadamia nut trees in the surrounding countryside. On the west coast, Kailua-Kona is warm, sunny, and sere, with wide vistas of blue Pacific water interrupted only by sparse grass and black volcanic rock. Just south of Kailua-Kona's airport is a small, climate-friendly power plant that operates like a giant battery, more formally known as ocean thermal energy conversion; electricity is generated by harnessing the difference in temperature between the ocean surface and seawater 1000 meters deep.[1] It's an uncommon way to harvest energy without generating greenhouse gases. More important for combatting climate change is another facility not far from Kailua-Kona. It's located on Mauna Loa, the biggest of the island's four volcanos, which run along the center of the island and stop rain clouds from reaching Kailua-Kona. On Mauna Loa is an observatory where scientists have been measuring levels of carbon dioxide in the atmosphere since 1958. It is home to the longest record of atmospheric carbon dioxide in the world, providing an unequivocal sign of how human activity is altering Earth's atmosphere with dire consequences for our climate.

Atmospheric carbon dioxide had been measured occasionally before, even back in the nineteenth century, but it wasn't until the 1950s when Charles David Keeling recognized the value of continuously monitoring this most important greenhouse gas.[2] Keeling realized he needed a pristine site when he found that carbon dioxide varied too much on top of Caltech's geology building in Pasadena, California, surrounded by gas-spewing cars and buildings, but also when surrounded by nature at Big Sur, in rain forests of the Olympic peninsula near Canada, or high mountain forests in Arizona near Mexico. Mauna Loa was the place he was looking for. Built by the U.S. Weather Bureau (now the National Weather Service) in 1956, the meteorological observatory on Mauna Loa is nearly ideal for keeping track of carbon dioxide and other greenhouse gases. It is far from crops and forests that consume carbon dioxide and far from cars, factories, and even tourists that produce the gas. Located on the north face of the volcano, the observatory is upwind from Kilauea, the youngest and most active of the volcanos on the Big Island, and at 3400 meters, it is high above local sources of carbon dioxide and other gases. About the same time as work started at Mauna Loa, Keeling also looked at carbon dioxide at Little America, a camp in Antarctica. Today there are about 70 ground-based monitoring sites around the world.[3] But Mauna Loa is iconic, the genesis of the Keeling Curve, showing the rise of atmospheric carbon dioxide from 1958 to today. The Curve is one of the most important graphs in environmental science (Fig. 1.1).

The Keeling Curve has been used to explore the twists and turns of carbon dioxide over 65 years and counting. Two of the most important conclusions from the Curve, however, were in Keeling's first report using Mauna Loa data.[4] Published in 1960, the report was the shortest of his career. Even at that early date, Keeling's data showed that carbon dioxide was increasing in the atmosphere by 0.5 to 1 parts per million (ppm) each year; today it is rising by over 2 ppm annually, from preindustrial levels of about 280 ppm to currently over 420 ppm. (A concentration of 420 ppm is equivalent to 420 molecules of carbon dioxide among a million air molecules, or 0.0420 percent.) Keeling

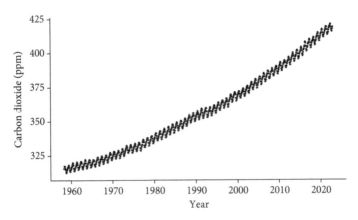

Figure 1.1 The Keeling Curve: concentrations of atmospheric carbon dioxide at the Mauna Loa Observatory. Data provided by Pieter Tans (NOAA Global Monitoring Laboratory) and Ralph Keeling (Scripps Institution of Oceanography), https://gml.noaa.gov/ccgg/trends/data.html.

concluded that the year-to-year increase was caused by humans, mainly our burning of fossil fuels, such as coal, oil, and natural gas. Keeling also observed that carbon dioxide levels didn't just change from year to year. More than a simple line sweeping up and up over time, the Keeling Curve is sawtooth, with highs in spring and lows in late summer. Keeling hypothesized that the seasonal pattern reflected the net release of carbon dioxide during winter and early spring when plant growth is slow, followed by net uptake in summer when plants are thriving, using carbon dioxide during photosynthesis.

Keeling didn't say anything about microbes. He mentioned land plants to explain why carbon dioxide swings up and down during the year at Moana Loa but not in Antarctica. Over six decades of additional data have confirmed that carbon dioxide changes more with the seasons in the Northern Hemisphere than in the South because the North has more land mass and land plants. Among many reasons why Keeling mentioned only land plants is that his short paper appeared about three decades before the importance of microbes in the biosphere was fully recognized. Since Keeling's 1960 paper, many publications have documented the participation of microbes in producing and consuming carbon dioxide and other greenhouse gases.

But no book, I suggest, has showcased the many connections between microbes and greenhouse gases, giving these unseen organisms a huge role in setting Earth's climate. Some books have explored how uptake and release of carbon dioxide and methane by microbes led to climate change in the geological past,[5] but we also need a book that explores what microbes are doing today if we are to fully understand the climate change that is now threatening the planet. Of course, we are the cause of this change, yet the complete story of how the burning of fossil fuels and other human activity is transforming the climate is more complicated and fascinating. It involves the unseen agents, the microbes.

What Is a Microbe?

Microbes are organisms that are visible only by microscopy. Collectively they have nothing in common except for their minute size. The Dutch draper Antonie van Leeuwenhoek first entered this unseen world in the 1670s when he trained his primitive microscope on samples from a nearby pond, rainwater, and scum from his teeth, and found what he called animalcules,[6] although many of Leeuwenhoek's "little animals" likely had plant-like properties. Modern microscopy has revealed a few defining characteristics of microbes, most notably the presence or absence of a nucleus, the membrane-enclosed organelle that houses a cell's genetic material. Eukaryotes have one, whereas prokaryotes, that is, bacteria and archaea, don't. Eukaryotes include the plants and animals we know from our macroscopic world, but they also include several inhabitants of the unseen world. Among the eukaryotic microbes are protists, such as some algae, protozoa, and amoeba—all single cells. Fungi are also eukaryotes. Even though many are microscopic and some like yeasts are single-celled, fungi are not usually classified with the protists.

Microbiologists have known about bacteria and protists since the nineteenth century or earlier. These microbes have been extensively studied because of the diseases they cause, such as cholera (the bacterium *Vibrio cholerae*), tuberculosis (another bacterium, *Mycobacterium tuberculosis*), and malaria (the protozoan *Plasmodium*). But microbiologists didn't know until the 1970s about the existence of another lifeform on Earth, archaea. Archaea were missed in part because they look like bacteria under the microscope, as both are prokaryotes and lack a nucleus. The difference between archaea and bacteria became clear only when Carl Woese and colleagues sequenced a key gene found in both.[7] In their 1977 paper, Woese and colleagues used "eubacteria" to denote the typical bacteria well-known to microbiologists for centuries and introduced "archaebacteria" for the other prokaryotes, represented at the time by methane-making microbes, the methanogens. Woese put "archae" in front of "bacteria" because he thought methanogens had a metabolism adapted to the environment presumed to exist on Earth during the Archean Eon three to four billion years ago when life started on the planet. Eventually, microbiologists dropped the "eu" from "eubacteria" and the "bacteria" from "archaebacteria" when the gulf between bacteria and archaea was fully appreciated. Microbiologists now know that although archaea were around in the Archaen, they aren't more ancient than bacteria. Archaea may have evolved from bacteria, and both came before eukaryotes.[8]

Microbiologists now recognize three domains of life: archaea, bacteria, and eukaryotes (Fig. 1.2). The first two are exclusively microbial while protists and many other eukaryotes are also too small to be seen without a microscope. The Tree of Life is populated mostly by microbes. Not being members of any domain of cellular life, viruses aren't on the Tree. Viruses aren't microbes, even though they are intensely studied by microbiologists and microbial ecologists. Arguably viruses are not even alive because they

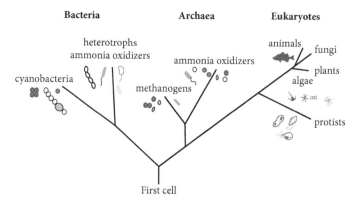

Figure 1.2 The Tree of Life, showing the taxonomic relationships among some of the microbes mentioned in this book. The complete Tree has many more branches and names. Most algae, which include the ancestor of plants, are single-cell eukaryotes, or protists. Except for animals and plants, all other organisms in the Tree are microbes.

are inert particles outside of the host cell they must infect in order to reproduce. Without their host cell, viruses have no metabolism and show no sign of life. Metabolic activity is key to how microbes affect greenhouse gases.

Unseen Agents

Thanks to their great abundance and diverse metabolisms, microbes are the agents that drive the natural cycles of carbon and of other elements in the biosphere. Microbes are the most abundant organisms on the planet and house more living material, or biomass, than any other organisms except for terrestrial plants.[9] Archaea, bacteria, fungi, and protists are ubiquitous and thrive in habitats where no other organism can. More than their numbers and mass, however, it's what they do, their metabolic activities, that makes them so important. Two of the most important activities are the production and consumption of carbon dioxide.

Carbon dioxide is produced by microbes and animals during the breakdown of the organic chemicals in the food consumed by these organisms. "Respiration" is one term for this type of metabolism, but "heterotrophy" is more informative, at least for a microbiologist. In heterotrophy, the organics provide not only energy but also the raw material for constructing an

organism's cellular constituents. The biochemical pathway by which heterotrophic bacteria, fungi, and other microbes release carbon dioxide is nearly
identical to the one found in animals. All these organisms, to use the language
of a chemist, oxidize organic chemical to carbon dioxide, with the electrons
from the organic going to oxygen gas. It's not essential to understand now
what's happening with these electrons, but following them becomes more
important later to understand more alien forms of metabolism. In any case,
the reaction for respiration can be represented like this:

$$CH_2O + O_2 \rightarrow CO_2 + H_2O$$

where CH_2O is shorthand for a generic organic chemical and O_2, CO_2, and
H_2O are oxygen (the gas), carbon dioxide, and water, respectively.

The reaction running opposite of heterotrophy is photosynthesis.
Microbes such as algae and cyanobacteria are capable of photosynthesis
like land plants and take up carbon dioxide. All these organisms are autotrophic because they can make their own cellular parts from carbon dioxide
and other nutrients, using the green pigment chlorophyll to capture the light
energy needed to run photosynthesis. Often co-occurring with algae are cyanobacteria, which were once thought to be a type of algae because they have
chlorophyll and carry out photosynthesis just like algae and plants. In fact,
they are in a very different domain of life, the Bacteria. Algae and cyanobacteria grow on exposed surfaces on land, evident from the green film covering
rock walls and the sides of houses, but they are most abundant and productive in lakes and the oceans.

The oxygen-producing version of photosynthesis, called oxygenic photosynthesis, carried out by plants, algae, and cyanobacteria can be represented
by an equation that is the reverse of the one for respiration, with the additional requirement of sunlight:

$$CO_2 + H_2O + light \rightarrow CH_2O + O_2$$

During oxygenic photosynthesis, light splits water to give the electrons
needed to reduce carbon dioxide to organic carbon, releasing oxygen gas
from the split water as a byproduct. The reduction of carbon dioxide to organic carbon, which requires energy, supplied here in the form of light, is the
reverse of the reaction that oxides organic carbon to carbon dioxide, which
releases energy. After millions of years when other types of photosynthesis

ruled, oxygenic photosynthesis evolved in cyanobacteria nearly 3 billion years ago or perhaps even earlier.[10] The first appearance of this form of photosynthesis on Earth is still unclear, but it was certainly around when oxygen became appreciable in the atmosphere about 2.3 billion years ago, a date so decisive that it's called the Great Oxidation Event. Before then, Earth didn't have much oxygen gas, and the biosphere was populated only by bacteria, archaea, and viruses for about the first half of the planet's existence.

Oxygenic photosynthesis is a great example of how microbes, working at the level of microns and molecules, have transfigured the planet on a global scale. The rise of oxygen remodeled the atmosphere, the ocean, rocks, and minerals. As oxygen increased, methane in the atmosphere and hydrogen sulfide in the ocean disappeared while iron on land turned into rust. Oxygen's appearance also sparked a revolution among the planet's biota, with a deadly outcome for most of the organisms that populated Earth for its first two billion years. Oxygen gas is dangerous. When dissolved in water, it forms free hydroxyl radicals, singlet oxygen, and hydrogen peroxide, which burn up a cell if left undefended. As oxygen levels rose after the Great Oxidation Event, many oxygen-sensitive organisms went extinct or were exiled to oxygen-free habitats such as organic-rich sediments or waterlogged soils. Yet the rise of oxygen was essential to the evolution of multicellular organisms, from dinosaurs and giant dragonflies to tiny shrews and primates, including *Homo sapiens*. By initiating the Great Oxidation Event, microbes, this time cyanobacteria, changed the course of the evolution and reshaped the biosphere.

Photosynthesis, with its consumption of carbon dioxide and release of oxygen, can now be put together with respiration, which produces carbon dioxide and consumes oxygen gas, into a simple version of the carbon cycle with just two pathways (Fig. 1.3). As a recent review put it, Earth's metabolism is run by a "solar-powered reduction-oxidation battery."[11] The battery is charged up by photosynthesis, which reduces carbon dioxide to organic matter with sunlight energy, and discharges when animals and heterotrophic microbes oxidize the organic matter back to carbon dioxide. A question to be explored in the next chapter is about the relative contribution of microbes to the two pathways. The short answer: a lot.

The real carbon cycle has many more pathways than just those given in Figure 1.3. Some will be discussed in this book. For nearly all, microbes have an important role, even when other organisms are more prominent. Plants are the main consumers of carbon dioxide on land, including about a quarter of the gas released by human activity,[12] yet plant growth depends

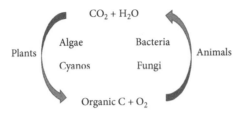

Figure 1.3 Two pathways in the carbon cycle and their connection to oxygen gas (O_2). Heterotrophic bacteria, fungi, and animals produce carbon dioxide (CO_2) during the degradation of organic matter ("Organics"), while algae, cyanobacteria ("Cyanos"), and plants consume carbon dioxide with the help of water (H_2O) during photosynthesis to produce organic chemicals and oxygen.

on microbes. Another carbon dioxide–consuming reaction, the weathering of silicate rocks, is usually viewed as being driven by purely chemical and geological forces, but it is modulated by microbes. The carbon cycle is mostly run by microbes as are the nitrogen cycle, the sulfur cycle, and all the other elemental cycles in the biosphere.

The Other Greenhouse Gases

Much of the book is about carbon dioxide, and even more could be said, but two other gases contribute a lot to the greenhouse effect. The next two most important greenhouse gases are methane and nitrous oxide. Both are increasing in the atmosphere, and both account for much global warming, even though their concentrations are lower than that of carbon dioxide. In 2022, the level of methane was 1.9 ppm and nitrous oxide's was about 0.3 ppm, much lower than the roughly 420 ppm of carbon dioxide.[13] Both methane and nitrous oxide contribute more to global warming than expected from their concentrations; methane and nitrous oxide account for about 23 percent and 6 percent, respectively, of the total anthropogenic effect.[14] (Carbon dioxide explains nearly all of the remaining 70 percent.) Methane and nitrous oxide are important because they are very effective in trapping heat in the atmosphere. Per molecule, methane is at least 20 times more effective than carbon dioxide, and nitrous oxide is even stronger, about 300 times more so than carbon dioxide over a 100-year time frame. Carbon dioxide is rightly first among anthropogenic greenhouse gases, but methane and nitrous oxide have oversized roles in setting the planet's heat budget.

Those numbers wouldn't be enough to include methane and nitrous oxide in this book, if not for their connections with microbes. The connections are much tighter than is the case for carbon dioxide. Animals and plants can produce or consume carbon dioxide, but they don't have much to do with methane and nitrous oxide. Both gases can be degraded by microbes and are produced only by microbes (not counting humans). The microbiology of methane is especially interesting, as Carl Woese demonstrated in his seminal studies. Although it can be degraded by bacteria, methane is produced mainly by the other type of prokaryote, archaea. (A few other organisms can make some methane, but archaea are the biggest producers.) The methane emitted by cows comes from archaea in their digestive tract. Other archaea and some bacteria release nitrous oxide. Although it is a distant third to carbon dioxide and methane in the ranking of greenhouse gases, nitrous oxide gains in prominence with its connections to other serious environmental problems. The common name for nitrous oxide is laughing gas, but there's nothing funny about the problems it causes or is connected to.

Another climate-relevant gas to be discussed here, dimethyl sulfide, is produced and consumed only by microbes in the oceans. Although it's not a greenhouse gas, dimethyl sulfide has a big, albeit controversial role in shading the planet and perhaps in preventing it from overheating. The controversial part comes from its connection to the Gaia hypothesis, named after the mother of all life in Greek mythology. According to the hypothesis, organisms shape the nonliving parts of the planet to maintain homeostasis and keep Earth habitable. Since its inception in the early 1970s, the hypothesis has been called "daring and provocative,"[15] and it has inspired decades of research and heated discussion that continues today. It has also been attacked as being anti-Darwinian, unprovable, and lacking a mechanistic basis. It even has been used by oil companies to argue that Earth is tough enough to withstand the impact of rising greenhouse gases.[16] Whatever the scientific merits of Gaia, dimethyl sulfide is a dramatic example of microbes shaping Earth's heat budget on a planetary scale.

To be complete, I need to say a few words about some greenhouse gases that have no direct connection to microbes. Water vapor accounts for the largest part, about 60 percent, of the natural greenhouse effect, by which I mean the warming of Earth by gases released by natural processes that would occur even in the absence of humans. Without water vapor and the greenhouse gases produced by microbes and other natural processes, the planet would be a freezing –18 degrees Celsius. The right amount of greenhouse gases is a good thing. The importance of water vapor has been seized on by climate skeptics who argue that carbon dioxide can't be a problem, and even atmospheric scientists in the early twentieth century believed that absorption of heat by carbon dioxide was insignificant compared to what water vapor does.[17] However, work published in the early 1950s demonstrated the large contribution of carbon dioxide to the greenhouse effect when it was discovered that the water vapor levels were high only close to Earth's surface.[18] The upper layers of the atmosphere are bone dry while carbon dioxide is relatively constant from the troposphere to the stratosphere. Also, carbon dioxide absorbs wavelengths of sunlight not absorbed by water vapor. Relevant to climate change is the fact that water vapor isn't increasing in the atmosphere like carbon dioxide, methane, and nitrous oxide are, and we don't alter water vapor on a global scale, except indirectly through our activities that produce greenhouse gases. Rather than water vapor causing global warming, the opposite is at work: global temperatures set levels of water vapor. Global warming by greenhouse gases released by human activity is changing the hydrological cycle by increasing evaporation and precipitation in many regions of the world.

Human activity definitely has lots to do with the other greenhouse gases worth noting here. These gases have one or more of the halogens chlorine, fluorine, and bromine, along with the usual carbon, hydrogen, and oxygen atoms. We use halogenated gases in a variety of applications, from refrigerants and aerosol propellants to insulation and fire extinguishers. Some, the chlorofluorocarbons (CFCs), are being phased out after it was discovered they destroy ozone, allowing DNA-damaging ultraviolet light to reach Earth's surface. The CFCs were replaced by another group of halogenated gases, hydrochlorofluorocarbons and hydrofluorocarbons, that don't touch ozone . . . but are potent greenhouse gases. Halogenated gases collectively account for 17 percent of the anthropogenic greenhouse effect.[19] Like methane and nitrous oxide, the concentrations of halogenated gases are low, roughly a million times lower than carbon dioxide levels, yet they pack a huge greenhouse wallop because they are so effective in trapping heat, and they last so long in the atmosphere. As one example, a molecule of sulfur hexafluoride lasts about 1000 years in the atmosphere and is over 18,000 times more potent as a greenhouse gas than a carbon dioxide molecule.[20]

Despite their importance, I won't say anything more about halogenated gases because unlike carbon dioxide, methane, and nitrous oxide, gases like sulfur hexafluoride are totally artificial and are not made by any natural process or organism besides humans. Worse, they are thought to be impervious to degradation by microbes. Even if there is some degradation, it's not enough to count for much.[21] Halogenated gases are exceptions to the rule that microbes produce and consume or degrade all chemicals found in the biosphere, most importantly the greenhouse gases carbon dioxide, methane, and nitrous oxide.

The Microbial Response

Microbes are not sitting still as climate change continues. Warming temperatures, droughts and downpours, more intense storms, wildfires, ocean acidification—all manifestations of climate change potentially have an impact on microbes. A key question to be addressed in this book is, how will microbes respond? A much longer book than this one is needed to thoroughly explore how climate change may alter the many things microbes do on Earth, from causing diseases in humans and the organisms we value to performing indispensable "ecosystem services" such as controlling the spread of pathogens and pests, providing nutrients to sustain plant growth, and purifying contaminated water, soil, and air. In this book, I'll mention some of these services, but I'll focus on how climate change affects what microbes do with greenhouse gases: will they consume more, or will they produce more as climate change continues and potentially intensifies in the coming years? The answers are crucial for predicting what faces Earth and its inhabitants in the coming decades.

Back on Mauna Loa, Keeling was able to make sense of why atmospheric carbon dioxide varied over the months and years without mentioning microbes, but by the end of this book, it should be clear that a complete understanding of climate change depends on knowing about microbes. Climate change, the biggest environmental problem now facing society, depends on the smallest, unseen organisms, the microbes.

2

The Natural Carbon Cycle and Fossil Fuels

A few years before the start of the Keeling Curve demonstrating the rise in atmospheric carbon dioxide, the Shell Chemical Company decided it needed to rebut a presentation by Edward Teller, the "father" of the hydrogen bomb, to the American Chemical Society in December 1957.[1] Speaking solemnly with a thick Hungarian accent, Teller warned that ongoing burning of fossil fuels would increase carbon dioxide levels in the atmosphere and raise global temperatures. A director at Shell, Marcus A. Matthews didn't dispute carbon dioxide's heat-absorbing capacity but argued there was nothing to worry about. The oceans could absorb the released carbon dioxide, and anyway, the amount of fossil fuels was small compared to what's in the natural carbon cycle. Matthews assured his readers that if atmospheric carbon dioxide did increase, it wouldn't be by much and its effect on the climate would be minimal. He was more concerned about society and industry running out of oil and other fuel reserves.

Decades of data showing the steady increase in carbon dioxide and its devastating effect on Earth's climate have proven Matthews spectacularly wrong. But he was right about the size of fossil fuel reservoirs, relative to the global carbon cycle. In some sense, they are small.

Matthews was mostly correct about what biogeochemists call stocks, or the amount of carbon present today as coal, oil, and natural gas, along with the carbon currently dissolved in the ocean and stored on land. He ignored, however, the other measure by which fossil fuels must be compared to the global carbon cycle, what biogeochemists call fluxes. For the carbon cycle, a flux is the amount of carbon being exchanged from one stock to another over a particular time, such as the uptake of carbon dioxide by a plant in a day or the release of carbon dioxide by a respiring animal each hour. A stock is like money in a savings account or stashed in a mattress, whereas a flux is like a salary that comes in every two weeks or the money that goes out daily or monthly to pay for food, rent, and electricity. Just as a thorough assessment of a person's finances depends on their savings and their salary relative to expenses, so too our understanding of the carbon cycle depends on both stocks and fluxes.

The amount of carbon in fossil fuels can be considered small only because other stocks of carbon are so big (Fig. 2.1). The Sixth Assessment Report (AR6) from the Intergovernmental Panel on Climate Change, better known by its acronym IPCC, says there are 1328 petagrams of carbon (Pg C) in coal, oil, and natural gas still left on the planet.[2] (A petagram is one million billon or 10^{15} grams, equivalent to a billon tons, or a gigaton, which is another unit commonly used by climate-change scientists.) That's not much compared to several other carbon stocks on the planet. One of the biggest is inorganic carbon dissolved in the oceans, 38,000 Pg C, followed by the 3100 Pg C in organic matter now locked in soils and permafrost. A large fraction, about 25 percent of all soil carbon, is in peat found mostly in the northern high latitudes.[3] The stock of soil carbon is double the amount of carbon in fossil fuels and three times bigger than what's in the atmosphere as carbon dioxide. Stocks of fossil fuel are even smaller than another natural stock, the carbon locked in sedimentary rocks from carbonate in the remains of organisms. This geological stock totals more than 150,000,000 Pg C, dwarfing all others.[4]

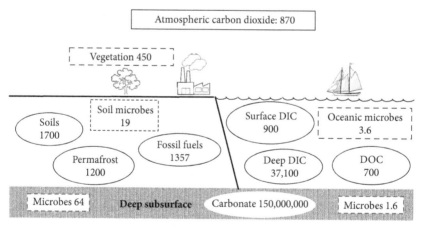

Figure 2.1 Carbon stocks in the natural carbon cycle and in fossil fuels. The numbers in the boxes and ovals are petagrams of carbon (Pg C). Two stocks in the ocean are dissolved inorganic carbon (DIC) and dissolved organic carbon (DOC). Organisms other than microbes and terrestrial plants are not shown because they have trivial amounts of carbon. The data for abiotic stocks are from IPCC AR6, whereas the data for microbes and vegetation are from Bar-On and colleagues (2018).

From the perspective of these parts of the carbon cycle, the amount of carbon in fossil fuels is small.

In the sections that follow, we'll go over natural fluxes of carbon dioxide— the release and uptake of carbon dioxide by microbes and other organisms— before turning to the carbon dioxide emitted during the burning of coal, petroleum, and natural gas. We'll see that like stocks, the flux out of fossil fuels as carbon dioxide is also small relative to what's happening in the natural carbon cycle. However, despite being small by some standards, the release of carbon dioxide from fossil fuel use and other human activity is having a huge impact on Earth's climate.

Carbon Dioxide from the Unseen Majority

One key to the power of microbes in driving the carbon cycle is their abundance and the amount of carbon in microbial cells, their biomass (Fig. 2.1). The first global estimate of biomass in one group of microbes was surprisingly high and attracted a lot of attention.[5] Barney Whitman and colleagues published a study in 1998 about the global stock size of prokaryotes, that is, bacteria and archaea. The title of the Whitman paper says it all: "The Prokaryotes: The Unseen Majority." The Whitman study used data on cell abundance in various habitats and the amount of carbon in an average cell to arrive at a number for the entire biosphere. They estimated that the planet has over 10^{31} prokaryotes, mostly bacteria, more than the number of sand grains on Earth (10^{19}) or stars in the observable universe (5×10^{22}), and that these prokaryotes contained as much as 550 Pg C. A decade later, another study argued that Whitman's numbers for deep habitats were too high by 10-fold and concluded that there were about "only" 10^{30} bacteria on the planet, containing 70 Pg C, mostly in deep subsurface environments.[6] Bacteria still constitute the majority in terms of population size, and even the lower numbers indicate that there is more carbon in bacteria and other microbes than in all other organisms on the planet, except for land plants (450 Pg C).

Impressive as those numbers are, they are not why microbes are indispensable in thinking about greenhouse gases. Fluxes are key. The size of the fluxes and how they vary have a huge impact on atmospheric carbon dioxide and thus Earth's climate. The flux of carbon dioxide into microbes occurs during photosynthesis, the physiological process behind primary production, while the flux out is respiration, mostly by heterotrophic microbes and animals.

Respiring organisms "burn" (a chemist would say "oxidize") organic chemicals made by photosynthesizing organisms back to carbon dioxide, just like
we burn fossil fuels and release that greenhouse gas.

On the land, much terrestrial respiration is by organisms in soil, about
half of the total (see later discussion). Soil respiration is estimated with a
chamber that traps carbon dioxide as it is emitted from the ground. In a study
published in 1927, the chamber was called a "respiration bell," although it was
shaped like an inverted funnel 27 centimeters in diameter at its wide end.[7] It
was made of zinc and coated inside to prevent absorption of carbon dioxide.
In the current version of the method, the chambers are plastic cylinders 10
to 20 centimeters in diameter, but the basic idea is the same.[8] The chamber
is placed on ground cleared of vegetation, with or without a collar inserted
into the soil, and the emitted carbon dioxide is measured every few seconds,
usually for a few minutes. A flow-through version compares carbon dioxide
levels in air flowing into the chamber versus the level flowing out after being
exposed to the soil. Although it is still common for someone to tend the
chamber in person, an automated version has made it possible to take many
more measurements, especially during inclement weather.

A study published in 2021 compiled all available results from the chamber
method and other approaches used at nearly 2500 sites scattered around the
world.[9] Most sites were in the Northern Hemisphere, home to two-thirds
of Earth's land, with many in North America, Europe, and China. The 2021
study reported that soils release as much as 101 Pg C of carbon dioxide each
year. Another publication that came out a year later said the average of all
studies is 87 Pg C per year.[10] Given all the possible errors and uncertainties,
the two estimates are remarkably close.

Still more work is needed to determine how much of soil respiration is
due to microbes. The publications cited here talk about "heterotrophic respiration" without mentioning the organisms, but nearly all of it is due to soil
microbes, mainly bacteria and fungi. Respiration by earthworms, mites,
nematodes, and other soil animals is small. The real challenge is separating
out heterotrophic respiration by microbes from respiration by roots. Soil
ecologists have tried to get estimates for the two types of organisms by physically removing roots and measuring respiration in rootless soils or by measuring respiration released by isolated roots and subtracting it from total
soil respiration. For forest soils, ecologists have girdled trees by removing a
ring of outer bark and phloem to stop the movement of organic chemicals
to the roots. After waiting a few months to ensure the roots are dead (and

eventually the entire tree), the ecologists measure soil respiration which by then is only by soil microbes. These approaches indicate that heterotrophic microbes account for more respiration than roots do in forest soils, but less in other soils; globally microbes account for 58 percent of soil respiration, roots the remaining 42 percent.[11] Microbes actually account for more and roots for less than those percentages indicate because root respiration includes respiration by symbiotic fungi living inside of or tightly bound to roots.[12] In any case, free-living fungi along with bacteria account for most of soil respiration outside of roots.

A study published in 2020 found that soil microbial respiration follows biological production, from highs in the Amazon and other tropical forests to lows in regions that are cold, arid, or both.[13] For 1980 through 2016, the yearly average for total soil microbial respiration was 57 Pg C, slightly more than the rate of 51 Pg C per year that I calculated from soil respiration and an assumed fraction (58 percent) of that respiration attributed to microbes.[14] In short, microbes account for a lot of respiration on land.

In the oceans the fraction of respiration carried out by microbes is even higher. In fact, they account for nearly all of it, as respiration by fish, dolphins, and even whales is trivial compared to what microbes do. The largest organisms in the oceans contribute the least to oceanic respiration because there are too few of them. As with soil fauna, large marine organisms have important roles in the carbon cycle but not in respiration. To figure out how much each type of microbe respires, microbial ecologists look at respiration in seawater after it has been passed through fine-meshed filters that separate organisms differing in size from 0.8 to 50 microns. A large part of oceanic respiration, roughly half, is carried out by the smallest microbes, the heterotrophic bacteria, in the less than 0.8-micron fraction.[15] Just as roots contribute much to soil respiration, marine microalgae and cyanobacteria also release a lot of carbon dioxide, what's called autotrophic respiration. This autotrophic respiration is less, as a fraction of total respiration, than that seen on land, but it still can be substantial, about 35 percent. Fungi account for much of respiration in soils but not much of it in the oceans. Small animal-like organisms, or zooplankton, at the edge of being visible to the naked eye, can release a lot of carbon dioxide and nutrients, but their respiration and the respiration of other nonmicrobial organisms bigger than 50 microns usually is a negligible fraction of the total.[16] Regardless of the specific type of organism, respiration in the oceans is microbial respiration.

Oceanographers face several problems when they try to estimate a respiration rate for the entire ocean. The ocean is vast, covering over two-thirds of the planet, deeper in parts than Mount Everest is high. Only waters close to shore are easily accessible. Both on land and in seawater, the carbon cycle varies over time, from day–night cycles to the comings and goings of the seasons, but in the ocean the carbon cycle also varies as currents, tides, and wind mix water horizontally over kilometers and vertically over meters. In addition to those challenges, the huge amount of dissolved carbon dioxide and its chemical relatives, bicarbonate and carbonate, makes it difficult for chemical oceanographers to see the changes necessary for estimating respiration and primary production. Instead, oceanographers often follow changes in dissolved oxygen over time. Carbon dioxide production and oxygen consumption are intimately connected, and rates of one can be converted to rates of the other.

One way to estimate oceanic respiration is like the chamber approach used by soil ecologists. Instead of chambers placed on the ground, oceanographers and aquatic ecologists put water into a small dark bottle and follow the decline in dissolved oxygen over a day or so. In addition to dark bottles, clear bottles, called "light bottles" in which both photosynthesis and respiration can occur during the day, are used to estimate net primary production, which is the difference between primary production and respiration by phytoplankton. The light–dark bottle method was first used in the 1910s by the Norwegians T. Gaarder, a chemist, and H. H. Gran, a botanist, as part of their attempts to understand primary production in the Oslo fjord.[17] Dissolved oxygen levels were measured by an even older method invented by Ludwig Wilhelm Winkler in 1888. Still used today, the Winkler method entails adding several chemicals to a water sample after it's been carefully enclosed in a special bottle designed to avoid introducing air. Even though the light–dark bottle method has been around for over a century, relatively few studies have used it in the oceans to estimate respiration. There are only a few thousand estimates of respiration from sites scattered across the ocean, at least 100 times fewer than the number of primary production estimates.

Only a couple of studies have reported respiration for the entire ocean. Paul del Giorgio and Carlos Duarte collected what data were available up to 2002 and estimated that respiration in the open oceans alone released 55 to 76 Pg C of carbon dioxide each year.[18] They thought respiration in coastal waters was substantial, but they didn't try to add anything to their open ocean number. Carol Robinson and Peter Williams did try to cover all marine waters and sediments and arrived at an estimate of 146 Pg C for a year,[19] which is even higher than total respiration on land. Most of the data used by these investigators came from the light–dark bottle approach. Unfortunately, there are several potential problems with the method. What happens in a few milliliters of seawater trapped in a bottle for a day may not represent what happens in a real ocean. Even if respiring microbes aren't perturbed by the bottle, it is challenging to do enough measurements with small bottles to cover all the planet's oceans over an entire year. One barrier is money, as always. The analyses must be done by a trained person on an oceanographic research vessel. Both are expensive, and access to research vessels is limited. As a result, the area of the ocean sampled for respiration is tiny, probably only a few thousand square kilometers out of the nearly 400 million square kilometers of the oceans, and the rates are measured only over a few days at best for a region.

A solution was to develop an approach that doesn't need bottles or ships. One bottle-free approach relies on following changes in dissolved oxygen levels over time in the aquatic habitat itself, that is, in situ, without incubating the seawater in a bottle. The in-situ change in oxygen is net community production, which is the difference between gross primary production and respiration. Oxygen levels typically increase during the day when primary production exceeds respiration, while at night levels decrease as respiration continues and photosynthesis shuts down. The daily changes in oxygen levels are followed with new oxygen sensors, such as optodes, that don't require seawater to be put into a bottle and can be deployed on smaller, unmanned, and thus cheaper vessels that can measure oxygen several times every minute for days without needing a break for meals or sleep. One type of "vessel," the Biogeochemical-Argo float,[20] rides currents passively and traverses the water column from the ocean surface to as deep as 2000 meters. The floats are part of the Argo fleet, named after the ship in Greek mythology, that Jason and the Argonauts sailed on in search of the Golden Fleece.[21] Jason is also an acronym for "Joint Altimetry Satellite Oceanography Network," a series of three satellites launched between 2001 and 2016 to measure sea surface height, which is used to explore rising sea levels, the speed and direction of currents, and heat stored in the oceans. Argo floats collect data only on temperature, salinity, and water pressure, while the biogeochemical version also looks at dissolved oxygen, dissolved nitrate (an important plant and algal nutrient), and a few other biogeochemical properties.

Ken Johnson and Mariana Bif at the Monterey Bay Aquarium Research Institute (MBARI) have used Biogeochemical-Argo float data to calculate rates of gross oxygen production and consumption in the global ocean.[22] MBARI is no stranger to robots and novel approaches. It has fostered the application of new technology for exploring the oceans ever since it started with support from David Packard,[23] one of the founders of the computer giant the Hewlett-Packard Company, known today as HP Inc. In the late 1980s, the David and Lucile Packard Foundation provided $13 million to establish MBARI, and since then has continued to support its marine scientists and engineers. Johnson and Bif used that support as well as other funds to analyze data from over 1000 Biogeochemical-Argo floats around the world from 2010 to 2019. Fluxes were based on day–night changes in dissolved oxygen (Fig. 2.2). Johnson and Bif had to assume that gross oxygen production, which is the total amount of oxygen released by photosynthesis before any consumption by respiring organisms, was equal to community respiration— the respiration by all organisms in the oceans. The assumption isn't entirely true, but it's necessary for using the float oxygen data to calculate rates, and Johnson and Bif convincingly argue the assumption doesn't affect their

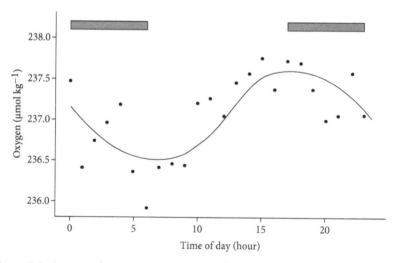

Figure 2.2 Average change in oxygen over a day in the oceans of the Northern Hemisphere (60° N to 10° N). The rectangles indicate night. The average dissolved oxygen concentration in equilibrium with the atmosphere (234 micromoles per kilogram of water) was added to the anomalies reported by Johnson and Bif (2021).

estimates. Although they focused on gross and net primary production, I'll use their numbers first to talk about respiration.

It is fairly straightforward to use float oxygen data to see how respiration varies with biological production and to estimate the amount of oxygen consumed by oceanic biota. Like what terrestrial ecologists had observed, Johnson and Bif's data imply that respiration follows production, with highs at the ocean surface and in spring when primary production is high and lows below the sunlight surface layer and in winter when primary production is low. The data also indicate that oceanic microbes with a bit of help from zooplankton consume about 380 petagrams of oxygen gas each year. The problem comes with converting petagrams of oxygen consumed to petagrams of carbon dioxide released. The basic equations for respiration and photosynthesis (Chapter 1) imply that for every molecule of oxygen consumed a molecule of carbon dioxide is produced; that is, the respiratory quotient is one. In fact, it varies a lot around an average of about 1.1.[24]

Using the 1.1 estimate, the float oxygen data imply that if 380 petagrams of oxygen are consumed, respiring organisms produce about 130 Pg C of carbon dioxide each year in the oceans. But that estimate may be too high. Some reactions produce and consume oxygen during photosynthesis without consuming or producing carbon dioxide. However, another study using float data and other approaches in the North Pacific Ocean found that oxygen consumption was roughly equal to carbon loss,[25] the ratio between the two ranging from 0.85 to 1.2. That ratio is about what one would expect for the respiratory quotient, suggesting that data on gross oxygen consumption give a reasonable estimate of how much carbon dioxide is released during respiration. In any case, the 130 Pg C calculated with Johnson and Bif's data is in fact lower than the 146 Pg C Robinson and Williams reported for global oceanic respiration using mainly light–dark bottle results. Still, all things considered, the two estimates are remarkably close, given the possible errors and the many ways things could go wrong.

Now we are ready to look at fossil fuels. One objective of looking at data from chambers, bottles, and floats is to compare natural fluxes, largely mediated by microbes, with a decidedly unnatural one, the release of carbon dioxide during fossil fuel consumption. Compared to the difficulties in figuring out natural fluxes, it is easy to estimate the carbon dioxide coming from fossil fuels. In 2020, carbon dioxide emissions from fossil fuels worldwide totaled 9.5 Pg C, slightly lower, by 5.4 percent, than the rate in 2019 before COVID-19 slowed everything down.[26] The pandemic also didn't much affect levels of atmospheric carbon dioxide. Anyway, after adding deforestation and other changes in land use that release carbon dioxide, the total flux from human activity was 10.2 Pg C in 2020.

That's a huge amount of carbon going into the atmosphere, but it's actually small compared to respiration on land and in the oceans. According to the numbers given before, Earth's biota release nearly 300 Pg C each year. Respiration by just microbes is nearly 200 Pg C every year, over 20 times emissions from fossil fuels and land use changes. It seems we should be much more concerned by the large natural flux of carbon dioxide emitted by microbes and other biota than by the much smaller flux associated with human activity. Yet the rise in atmospheric carbon dioxide is well-known to be caused by what we humans are doing. Where is all that carbon dioxide from respiring microbes and other organisms going? The short answer is, it's going into photosynthesizing organisms, including many microbes.

Global Primary Production and the Smallest Photoautotroph

Primary production is one of a few processes on Earth that take carbon dioxide out of the atmosphere. It is biology's answer to respiration, the yin and yang of the carbon cycle. As with global respiration, estimating global primary production usually means getting an estimate for land and another for the ocean. The experimental approach for the two realms differs, as does the contribution by microbes. On land, nearly all primary production is by trees, grasses, and shrubs: vascular plants, not algae or cyanobacteria. Primary production by microbes is crucial for supporting food webs in lakes, ponds, and rivers, but the amount of carbon dioxide consumed by these freshwater microbes is dwarfed by what terrestrial plants do.

The rate of terrestrial primary production has been estimated by following carbon dioxide in the air from instrumented towers ranging in height from a few meters above grasslands to several tens of meters above a forest floor. Carbon dioxide decreases during the day when photosynthesis is active and increases at night when only respiration operates. Net community production is estimated by a statistical approach called eddy covariance.[27] After the first tower was put up in the late 1990s, there are now over 900 eddy covariance sites around the world, together making up the FLUXNET network. According to a study published in 2019 that combined FLUXNET data and a satellite approach,[28] yearly gross primary production on land is 147 Pg C. That's higher than the estimate (113 Pg C per year) of all values published up to 2022, but it's close to the estimate (149 Pg C per year) derived by a study using respiration data.[29]

According to the numbers just cited, terrestrial primary production is more than enough to consume all the carbon dioxide released by respiring organisms on land each year (Fig. 2.3). The 51 Pg C released annually by soil microbes is more than matched by land plant primary production. What about the oceans? It's a bit more complicated.

Unlike on land, nearly all primary production in the oceans is done by microbes. Seaweeds and sea grasses provide valuable food and shelter in nearshore habitats, but they cover too small of an area to contribute much to global estimates. The main primary producers in the oceans and other aquatic habitats are free-floating photosynthesizing microbes, or phytoplankton, the plant-like counterpart to zooplankton. Phytoplankton include algae like diatoms and coccolithophores, which are eukaryotes that vary from single cells a few microns in diameter to chains or complex colonies of cells hundreds of microns long. Phytoplankton also include cyanobacteria, which are prokaryotes that carry out oxygenic photosynthesis like algae and plants do. Although some marine cyanobacteria make long chains big enough to

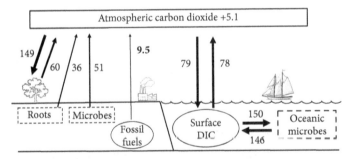

Figure 2.3 Natural fluxes of carbon dioxide and release by fossil fuel burning. All numbers are Pg C per year. DIC is dissolved inorganic carbon. Estimates for the increase in atmospheric carbon dioxide (+5.1 Pg C per year) and the atmosphere-ocean fluxes are from IPCC AR6. For the ocean, the gross primary production estimate is from Huang and colleagues (2021), whereas the respiration estimate is from Robinson and Williams (2005). The flux estimates for terrestrial biota are from Jian and colleagues (2022), assuming roots account for 42 percent of soil respiration and aboveground foliage is 62 percent of autotrophic plant respiration.

be visible to the naked eye, the most abundant types in the ocean are small, single cells, only about one micron or less in diameter. Having a small cell is an advantage for taking up dissolved nutrients when concentrations are low, such as in the open oceans far from land. Big phytoplankton, 20 microns or larger, are abundant only in nutrient-rich seas. Because nutrient-poor waters in the open oceans cover so much area, about half of all marine primary production is by the smallest phytoplankton, the cyanobacteria.

One genus of marine cyanobacteria, *Prochlorococcus*, has fascinated biological oceanographers and microbiologists since its discovery in the late 1980s by Sallie "Penny" Chisholm,[30] with key contributions from Rob Olsen who was a postdoc in Chisolm's lab at the time. Growing up in Marquette in the Upper Peninsula of Michigan, Lake Superior was Chisholm's ocean.[31] She got to know a real ocean when she did a postdoc at the Scripps Institution of Oceanography located in southern California, after which in 1976 she joined the faculty at the Massachusetts Institute of Technology. During the early part of her career, Chisholm studied cell division of diatoms using a laser-based instrument, called a flow cytometer. It was designed for counting red blood cells, not for looking at marine microbes. Chisholm and Olsen stumbled on *Prochlorococcus* when they took a flow cytometer out to sea to look for another abundant genus of marine cyanobacteria, *Synechococcus*. In the North Atlantic Ocean, they noticed cells that looked somewhat like *Synechococcus* but were smaller with a different color.[32] Sequences of the 16S rRNA gene, which is commonly used to classify bacteria, eventually revealed that the smaller cells were related to *Synechococcus* but were different enough to warrant their own genus name, *Prochlorococcus*. In addition to the sequence data and cell size, *Prochlorococcus* has different photosynthetic pigments than those found in marine *Synechococcus* and other cyanobacteria. The distribution of the two cyanobacterial genera among oceanic waters differs, as do other aspects of their ecology.

Decades of work by Chisholm and others have confirmed the importance of *Prochlorococcus* in the oceans and for understanding primary production on the planet. Among the many types of phytoplankton and the huge diversity of life in the oceans, *Prochlorococcus* stands out, first because it is very abundant, reaching densities of 100,000 in a milliliter of seawater. It is the most abundant photoautotroph on the planet: much more abundant than the photoautotrophs on land—that is, plants—and often more abundant than algae and other cyanobacteria in the oceans.[33] (See Box 2.1 for a definition of photoautotroph.) *Prochlorococcus* is not only the most abundant

---

Box 2.1 What Is a Photoautotroph?

Microbiologists and microbial ecologists use the term to describe organisms that get their carbon from carbon dioxide (they are autotrophs) using light energy (hence "photo"). It's a more precise, informative term than "plant-like." Plants, algae, cyanobacteria, and other types of photosynthetic bacteria are all photoautotrophs.

---

photoautotroph; it's also the smallest one, about 0.6 microns in diameter, even smaller than *Synechococcus* with its 1.0-micron cell. It also has the smallest genome of any photoautotroph. Some isolates have only about 1700 genes in a genome with 1.65 million base-pairs, a fraction of the 10,972 genes and 14 million base-pairs genome in *Scytonema*, a large mat-forming cyanobacterium that has the biggest genome of all cyanobacteria.[34] Both are tiny compared to the genomes of plants and animals. The black cottonwood (*Populus trichocarpa*), for example, has about 73,000 genes in a 485 million base-pair genome,[35] while the human genome has only about 20,000 genes and 3000 million base-pairs. (The numbers of genes just given are for those for making or coding for proteins.) With that small cell, directed by that tiny genome, *Prochlorococcus* manages to account for roughly half of phytoplankton biomass and production in the ocean.

The question is whether this small cyanobacterium and the other phytoplankton produce enough oxygen and consume enough carbon dioxide via photosynthesis to offset the huge amount of respiration in the ocean. One way to answer the question is to compare gross primary production with total respiration. Unfortunately, there are about as many estimates of gross primary production as there are of total respiration, which is to say, not many. One study published in 2021 by Yibin Huang and colleagues[36] estimated gross primary production by first compiling data from light–dark bottles and the triple isotope method, which uses the three nonradioactive, or stable isotopes of oxygen: ^{16}O (99.76 percent of all oxygen atoms), ^{17}O (0.04 percent), and ^{18}O (0.20 percent). The triple isotope method takes advantage of the fact that the abundance of ^{17}O relative to the two other isotopes is lowered by photochemical reactions in the stratosphere while it increases in water because of photosynthesis. Thus, the relative level of ^{17}O in seawater can be used to estimate primary production without incubating seawater in bottles.

The 2021 study then turned to another source of data used by many earth scientists who want to get estimates for the entire planet: satellites. Biological oceanographers started to use satellites to look at the color of the surface ocean in the late 1970s.[37] Key was figuring out how to detect the light reflected by the phytoplankton pigment chlorophyll buried amid the large background of light reflected to the satellite by air and seawater. That light reflected from chlorophyll is then used to deduce its concentration and in turn the level of phytoplankton biomass. Additional uncertainties and potential errors arise in converting a chlorophyll concentration to phytoplankton biomass, because the amount of the pigment in a phytoplankton cell varies among phytoplankton species and light intensity (cells have more chlorophyll under low light). Despite the problems, chlorophyll remains an invaluable measure of phytoplankton in the ocean and also of plant biomass on land, in no small part because it can be seen from space.

The first test of satellite oceanography came with the launch in October 1978 of the Coastal Zone Color Scanner Experiment (CZCS), which according to Howard Gordon, one of the pioneers of the field, was considered by many to be a "boondoggle being carried out by lunatics."[38] Soon after its launch, Gordon and others sailed off to measure chlorophyll in surface waters off Florida's west coast in the Gulf of Mexico when the satellite passed overhead, so they could compare the satellite numbers to real chlorophyll levels. One pass of the satellite, Orbit 296, proved especially informative. Gordon's and his colleagues originally planned that their ship, the Athena II, a 165-foot decommissioned patrol boat from the Vietnam War, would steam on a straight line from Key West to Tampa, Florida. Instead, the ship's crew veered off course to get closer to shore so that they could get better TV reception and watch Monday Night Football. With the detour, as the crew watched the Raiders beat the Bengals 34 to 21, the ship fortuitously went through a phytoplankton bloom with high chlorophyll concentrations. It made for a great test of the Coastal Zone Color Scanner and was an important step in showing the power of satellites in assessing phytoplankton over vast expanses of coastal and open oceans. Satellites have been indispensable in oceanography and all climate-change science ever since.

The 2021 study by Huang and colleagues used a machine-learning approach that combined satellite data with light–dark bottle and triple isotope data to estimate global rates of gross primary production in the ocean. They found it to be somewhere between 103 and 150 Pg C per year. The estimate from the Johnson and Bif study using Biogeochemical-Argo float data, 130

Pg C per year, is in the middle of this range. The high end is just enough to take care of the 146 Pg C estimated to be respired by oceanic microbes, but the low end indicates a potential problem. Perhaps respiration somehow exceeds gross primary production in the ocean. Perhaps the ocean's metabolism is out of whack.

The metabolic state of the ocean was a big controversy about 20 years ago and still simmers today. The controversy was launched by del Giorgio and Duarte, the two ecologists who provided some of the first estimates of global respiration in the oceans. Using mainly data from the light–dark approach and others that rely on incubations, Duarte has championed the view that respiring microbes produce more carbon dioxide than is consumed by primary production in vast regions of the open oceans.[39] These regions, covering 85 percent of the oceans, are net heterotrophic according to Duarte, with respiration exceeding primary production. If Duarte is correct, it has big implications for understanding food webs and for looking to the oceans for sequestering carbon dioxide away from the atmosphere. If respiration exceeds oceanic primary production, Duarte must find other sources of organic carbon to support so much respiration. He and his colleagues argued that the heterotrophic regions are supported by carbon transported from highly productive regions of the ocean, those that are net autotrophic, and by organic carbon from the atmosphere. Many have argued against Duarte and his supporters. Biological and chemical oceanographers have pointed out that these outside sources of carbon from autotrophic regions are not big enough to support heterotrophic regions, and they identified problems using incubations in bottles to estimate primary production.[40] In-situ-based approaches, such as following dissolved oxygen over time and using stable carbon isotopes, give higher estimates of primary production than do approaches relying on bottle incubations. Studies using in-situ and stable isotope approaches indicate that primary production usually exceeds respiration in nearly all oceanic regions.

So those data indicate that there is enough primary production by microbes to consume the carbon dioxide produced by microbes in the global ocean. Although many questions remain unanswered, I believe it's fair to say that most oceanographers today think the oceans are mostly autotrophic with primary production exceeding respiration, allowing for the extra organic carbon to be stored in the water column and sediments. That's good news for the rest of the marine food chain and the capacity of the oceans to sequester carbon away from the atmosphere.

To recap, in the ocean as well as on land, more carbon dioxide is consumed than is produced by microbes, making possible the storage of carbon and the buildup of huge pools in both ecosystems: soil organic matter on land and dissolved organic and inorganic carbon in the ocean. A biogeochemist or climate scientist would say that land and the ocean are sinks for carbon dioxide—the two ecosystems are taking up more carbon dioxide from the atmosphere than they are releasing. Another takeaway from all the numbers just presented is that primary production and respiration are about same on land and in the sea. We know more about net primary production on land versus the ocean; as mentioned before, net primary production is gross primary production minus respiration by plants and phytoplankton. The annual rate of net primary production on land is about 56 Pg C, only a bit higher than the 48 Pg C for the oceans,[41] despite vast differences in the organisms responsible for this production. Trees, shrubs, and grasses are much larger than the algae and cyanobacteria carrying out primary production in the oceans, and the amount of carbon in terrestrial vegetation, about 450 Pg C, is several hundred times the amount of carbon in phytoplankton.[42] Tiny cyanobacteria and algae can carry out as much primary production as giant redwoods and sequoia do because microbes are so abundant and live in the biggest habitat on the planet, and because they grow very fast. Phytoplankton double in numbers about every day, while fast-growing weeds need weeks, and trees decades, to reach a reproductive age. The rate of primary production on land and in the oceans is similar, but the organisms carrying it out couldn't be more different.

All these numbers are important in thinking about carbon dioxide levels in the atmosphere and so also about climate. Even though carbon dioxide consumption by primary producers is greater than its production by respiration on land and in the oceans, the difference isn't huge, a few petagrams of carbon each year. The small difference is a sign that the global carbon cycle is currently in balance, and other evidence indicates it has been so, more or less, for the last 10,000 years, when the planet came out of the last ice age. At the height of the last ice age, about 20,000 years ago, atmospheric carbon dioxide was only 180 ppm, rising to 280 ppm when the ice finally receded. It stayed at that level even during the Medieval Warm Period, between 750 and 1350 CE, when England was warm enough for vineyards and Greenland was green enough to encourage Viking settlements.[43] Atmospheric carbon dioxide remained at 280 ppm until the start of the Industrial Revolution in the nineteenth century. Today it's about 420 ppm.

After 10,000 years of stability, the natural carbon cycle is now being disrupted by human activity. Even though the release of carbon dioxide from fossil fuels and other human activity is small compared to global primary production and respiration, it is large compared to the slight imbalance between the two natural fluxes. It's why carbon dioxide is increasing in the atmosphere. The question then becomes, how will the natural carbon cycle respond? What will microbes do? As the next few chapters will discuss, there is some good news on land and perhaps not so good news from the ocean.

3

Carbon Sinks and Sources on Land

Plants seem to dominate the carbon cycle on land. After all, more biomass is in grasses, shrubs, and trees than in any other organism, more than what is in elephants, hippos, and whales, even more than what's in microbes (Chapter 2). More important than biomass is the contribution by plants to a crucial flux in the carbon cycle: consumption of carbon dioxide during photosynthesis. Plants are the reason why land is a big sink of carbon dioxide, soaking up so much of the greenhouse gas from the atmosphere, including what is released by fossil fuel burning and other human activity. According to a 2022 status report of the global carbon cycle, of the roughly 10 petagrams of carbon (Pg C) released as carbon dioxide each year by humans, land plants take up about 29 percent, slightly more than the ocean's 26 percent, leaving 5.1 Pg C to accumulate in the atmosphere every year.[1] Without land plants, global warming would be even worse than it now is.

But plants aren't doing it alone. The biggest organisms on land, responsible for one of the largest sinks of atmospheric carbon dioxide on Earth, depend on the smallest organisms, the microbes. Giant sequoias and towering Douglas fir trees couldn't take up so much carbon dioxide without help from microbes. One well-known way in which microbes help is in supplying the nutrients needed by land plants to carry out photosynthesis and to grow and reproduce. The growth of trees, shrubs, and grasses in the absence of artificial fertilizers is sustained by nitrogen and phosphorus nutrients released during the decomposition of detritus by soil microbes. Plants have many even more intimate relationships with microbes. Legumes like peas and beans get their nitrogen from symbiotic nitrogen-fixing bacteria that live in cellular structures called root nodules. Formed after the root and symbiotic bacterium exchange complex chemical cues, root nodules are evidence of a highly evolved symbiosis between plants and bacteria.

Even more common is the relationship most plants have with fungi. The symbiotic relationship between plants and fungi is key to understanding the terrestrial carbon cycle and its response to climate change.

Fungal Connections

Plant–fungus relationships are ancient, dating to the emergence of both from aquatic habitats and their appearance on land.[2] By the time plants had evolved from an aquatic green alga and ventured onshore, fungi were already there, having escaped their aquatic past as a flagellated protozoan.[3] Important groups of fungi had evolved far enough to help early plants colonize land about 450 million years ago, although at first fungi may have been more adversarial than helpful, according to genomic analyses of primitive plants, which have documented how fungi evolved from pathogens to mutualistic partners with their plant hosts over millions of years. Plants needed the help because the early days on land were tough. Without a well-developed root system, the first terrestrial plants faced challenges in securing nutrients and water from the barren landscape of the early Paleozoic Era. Ever since then, the partnership continues to benefit both fungi and plants and has contributed to the high abundance and huge diversity of both on land.

The most important fungus–plant partnership takes place on roots (Fig. 3.1). The fungal partners are known as mycorrhiza, from the Greek for fungi and root. Those that make a bush-like structure after penetrating plant root cells are called arbuscular mycorrhiza, while ectomycorrhizal fungi form a sheath around the root tip. In either case, fungal cells strung together—the hyphae—grow away from the root and fan out into the soil to make an underground network called a mycelium. The fungus provides several benefits to its plant host. It can act as surrogate roots, taking up nutrients and water

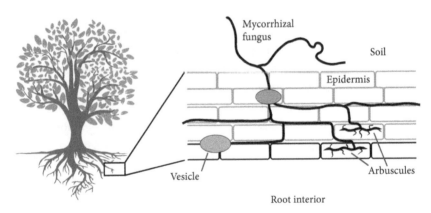

Figure 3.1 Mycorrhizal fungus colonizing a plant root. The right part of the diagram is a close-up view of the fungus interacting with outer cells of the root.

in interstitial spaces of soils where roots are too big to venture. The fungus then ferries these essential nutrients back to plant roots via the mycelium network, and in exchange the plant provides sugars needed by the fungus for its growth. In effect, the plant has outsourced some of its root functions to mycorrhizal fungi. As evidence of the importance of this outsourcing, over 90 percent of plant species are host to some sort of mycorrhizal fungi, and plants may transfer as much as 30 percent of what they make during photosynthesis to their mycorrhizal symbionts.[4] Equipped with these fungi, plants have conquered land and now are a large sink for atmospheric carbon dioxide.

Other fungi, known as saprotrophs, aren't symbionts with plants but rather earn their living by decomposing plant detritus, especially wood. The organic chemical that defines wood, lignin, is a very large, complex molecule consisting of phenol-like, or aromatic components, made by the phenylpropanoid pathway (Fig. 3.2).[5] Although phenol may not be familiar, readers would recognize the odor of some its derivatives in vanilla, bacon, and even in the bouquet of some wines.[6] (The fragrance of these chemicals is the reason the term "aromatic" was first used to describe them, although chemists now call a chemical aromatic when it has a certain kind of ring made of carbon atoms regardless of its odor.) The first land plants used products from the phenylpropanoid pathway for protection against DNA-damaging ultraviolet light. When cell walls acquired phenylpropanoids via evolution and became lignin, plants were better equipped to withstand desiccation and attack by herbivores and pathogens. The stiffness of lignin also enabled plants to grow upright, eventually tens of meters aboveground, supporting tens to hundreds of tons of leaves and branches. Because of lignin, wood is tough, by some measures stronger than steel. Lignin also makes wood difficult to degrade and decompose. Although bacteria degrade some lignin in the absence of oxygen, lignin is mostly decomposed by a few saprotrophic fungi when oxygen is sufficient. Decomposition of wood and other plant detritus is nearly 100 percent efficient. Virtually everything made by a plant eventually is turned back to nitrogen and phosphorus nutrients and carbon dioxide.

This is generally true, except for one crucial period in the geological past when microbes fell behind and plant debris built up. Our climate is now threatened as a result. The problem was caused by lignin-rich trees that evolved during the Devonian period when the diversity of trees and other plants exploded, impacting all biogeochemical cycles, including that of carbon.[7] By the mid-Devonian about 390 million years ago, a tree-like

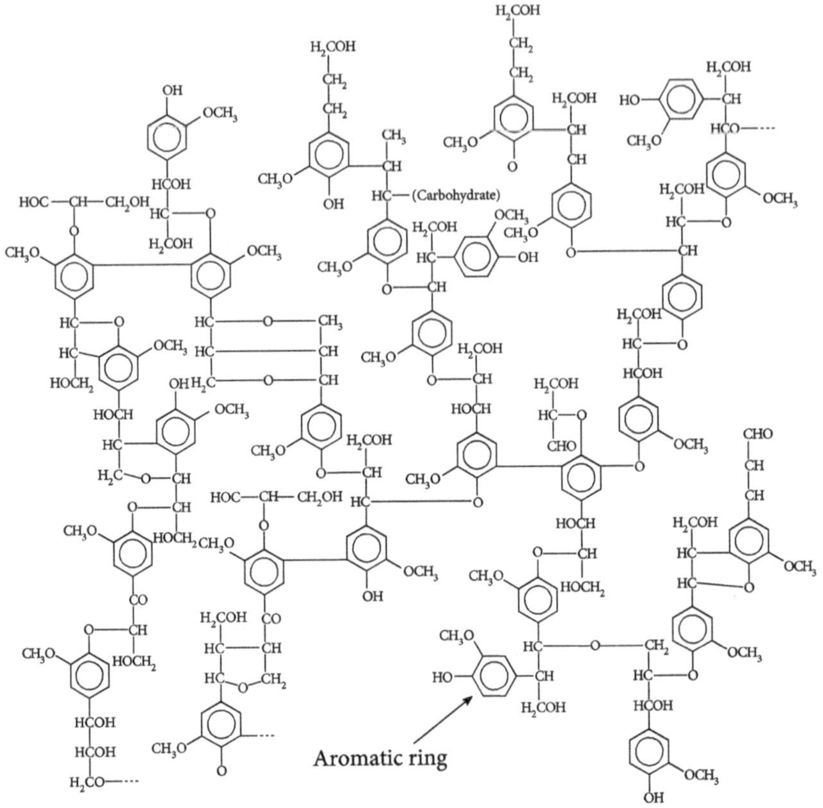

Figure 3.2 One model of lignin, a major component of wood. The arrow points out only one of the many aromatic rings in lignin. From Karol Głąb, licensed under the Creative Commons Attribution-Share Alike 3.0 Unported.

plant had evolved enough to make a forest, known today by its fossilized remains in the Riverside Quarry near Gilboa, New York.[8] This plant didn't have leaves and instead carried out photosynthesis in branch-like structures. The first plant recognizable as a tree, called *Archaeopteris*, had an extensive root system like modern trees and grew to over 20 meters. That was nothing compared to another primitive tree related to clubmosses which grew to over 30 meters.[9] About 300 million years ago, clubmosses along with tree ferns and great horsetails flourished in vast low-lying swamps while giant centipedes, cockroaches, and scorpions scurried in the understory, and dragonflies as big as birds flew overhead.[10] This was the Carboniferous, a term derived from the Latin for "coal bearing." From 359 million years ago to the start of the

Permian 299 million years ago, wood and other detritus from primitive trees were buried in swamp sediments and eventually were transformed by geological forces into coal. Because saprotrophic fungi and bacteria didn't keep up with growth by primitive vegetation, dead plant remains accumulated in the muck of the Carboniferous swamps and eventually turned into the dirtiest of all fossil fuels. (Burning coal produces more carbon dioxide than oil or natural gas does while also releasing heavy metals, fine particulates, and other deadly pollutants.)

Why couldn't microbes keep up? An early hypothesis focused on lignin and its decomposition by fungi.[11] The fungi extant at the time were hypothesized to be ineffective in decomposing detritus from primitive vegetation, because evolution of lignin-degrading enzymes in fungi lagged evolution of lignin synthesis in early trees. In support of the lag-in-evolution hypothesis, an analysis of 31 fungal genomes suggested that key lignin-degrading enzymes evolved too late to prevent the massive accumulation of lignin-rich plant detritus.[12] The enzymes finally appeared near the end of the Carboniferous, coinciding with a decline in the burial of plant organic carbon. According to this hypothesis, lignin-degrading enzymes were not around to stop the buildup of plant detritus that went on to form coal.

There are several problems, however, with blaming evolution. Other genomic and fossil evidence indicated fungi related to known lignin-degraders with important lignin-degrading enzymes had evolved by the early Carboniferous. Enough fungi with efficient lignin-degrading enzymes were around and should have stopped coal formation. They were certainly around later, in the Cretaceous–Paleogene roughly only a million years ago, yet coal also formed during that time in western North America. The Cretaceous–Paleogene coal beds are smaller than those formed during the Carboniferous for other reasons, not because the evolution of lignin degradation had finally caught up to lignin synthesis by plants. If microbes with the right enzymes were around, why then did so much coal form during the Carboniferous?

Geologists now think that the extensive coal beds of the Carboniferous came about when climate and geologic forces combined to create vast areas with the right physical condition for preventing the efficient degradation of plant detritus.[13] The right condition was the lack of oxygen, or anoxia. Oxygen was stripped from the muck of Carboniferous swamps as they filled with dead leaves and branches from clubmosses and tree ferns. Although degradation of most organic chemicals is slower in anoxia, lignin degradation is especially sensitive to oxygen levels. White-rot fungi, the most powerful lignin-degrading microbe, need oxygen or reactive chemicals made from oxygen like hydrogen peroxide to break up the many aromatic rings found in lignin.[14] This biochemical necessity for oxygen means fungi were fated to fall behind and had no chance to prevent the formation of coal regardless of their efficiency in decomposing plant detritus in oxygen-rich environments.

The burial of plant detritus and the formation of coal beds radically altered the planet's atmosphere and its climate as the Carboniferous period drew to a close 299 million years ago. Microbes, especially fungi, played a role, if only in what they failed to do. Degradation of all the plant detritus left behind during the Carboniferous would have used a lot of oxygen gas and returned an equally large amount of carbon dioxide to the atmosphere. Instead, because so much organic matter went undegraded by microbes, oxygen gas levels in the atmosphere soared to as high as 35 percent,[15] much higher than today's 21 percent. Atmospheric oxygen diffuses too slowly into thick mud to help with detritus degradation, yet it happens fast enough in air to enable the evolution of bird-sized insects during the Carboniferous. The high oxygen levels fueled extensive forest fires that left behind charcoal in the fossil record. While oxygen increased to record highs, atmospheric carbon dioxide decreased to near-record lows.[16] When carbon dioxide levels dropped and the greenhouse effect weakened, Earth became cooler and drier, and large regions were covered in ice sheets, ice caps, and glaciers. This was the Late Paleozoic Ice Age, which started about 340 million years ago and lasted roughly 55 million years. It is ironic that the formation of coal beds created this climate change millions of years ago while today the burning of that coal is a major reason why Earth is now threatened by climate change.

It would be oversimplistic to ascribe all climate change during the late Paleozoic to land plants and fungi. The complete explanation involves, among several factors, plate tectonics, weathering (which is connected to plants and mycorrhizal fungi—see Chapter 6), and variation in the amount of sunlight reaching the Earth, what climatologists call Milankovitch cycles. Still, thanks to help from mycorrhizal fungi, plants did colonize land and went on to produce a lot of organic matter during the Carboniferous. If saprotrophic fungi had somehow evolved a mechanism to degrade lignin and other plant detritus in the absence of oxygen, much less coal would have formed. It seems fungi and other microbes deserve a bit of blame. The extent to which they can be blamed points to the importance of what microbes do (or not do) in effecting climate change.

Regardless of what happened millions of years ago, there is no doubt that today microbes are essential to the terrestrial carbon cycle and its response to climate change. Mycorrhizal fungi, nitrogen-fixing bacteria, and nutrient-releasing microbes of all sorts enable terrestrial primary production to be the biggest sink for atmospheric carbon dioxide on the planet while other terrestrial microbes are responsible for returning much of that carbon dioxide back to the atmosphere. The balance between these two carbon dioxide fluxes determines how much of this potent greenhouse gas is emitted from terrestrial habitats and how much carbon is left behind in soils.

The Enigma of Soil Organic Matter

A lot of carbon has been left behind even in well-aerated soils, and has built up over eons to form one of the largest stores of carbon on Earth, known as soil organic matter (SOM). Soil organics contain over five times more carbon than is in coal,[17] and there is even more carbon stored in soil than is present in terrestrial vegetation or in the atmosphere. Organic matter in soils is one of the largest stocks of carbon on the planet, second only to the inorganic carbon dissolved in the ocean or tied up in rocks. You may think such a large thing would be easy to study, but that is far from the case. Although soil chemists have worked on the stuff for over 200 years, only recently have they begun to understand the chemical nature of SOM and its origin.[18] The enigma is now being solved.

For over a century, soil scientists thought that most SOM came about by "humification," a chemical process that transforms simple, easily degraded cellular components made by plants into hard-to-degrade, or refractory chemicals called humic substances, found in humus, the dark, earthy stuff of rich soils. Humic substances were defined by how they were isolated from soils, which involves harsh extractions with acids and alkaline solutions. These substances were thought to be large, complex macromolecules that are resistant to degradation by microbes, so they were thought to make up much of SOM. The humification hypothesis seemed reasonable. It explains how sugars, starch, proteins, and other cellular components of plants that are easily degraded can turn into organic matter that lasts for thousands of years.[19] The age of organic matter can be deduced from how much of a radioactive isotope of carbon, ^{14}C, it contains (Box 3.1). Armed with the ^{14}C-dating method, soil scientists have estimated that organic matter at the soil

---

Box 3.1 Carbon Dating

All natural organic chemicals have trace amounts of ^{14}C put there by photosynthetic organisms assimilating naturally occurring ^{14}C-carbon dioxide made by cosmic rays acting on nitrogen gas in the atmosphere. Plants, algae, and cyanobacteria take up ^{14}C-carbon dioxide along with ^{12}C-carbon dioxide, the stable (nonradioactive) and most abundant isotope of carbon. Once the organic is synthesized, no new ^{14}C can be added, and what's there starts to decrease at a known rate, decaying to a nitrogen atom. The amount of ^{14}C relative to ^{12}C is used to deduce the organic matter's age.

---

surface is a thousand years old on average, increasing to 10,000 years or older below the surface. The thousand-year-old organic chemicals must be refractory, it seemed, the end products of a chemical process that microbes couldn't touch. The humification hypothesis seemed to explain the enigma of soil organic carbon: how can it exist when so many biological forces are out to destroy it?

Arguments against the humification hypothesis were made soon after it was proposed. One early skeptic was Selman Waksman, a soil microbiologist who emigrated from Ukraine to the United States in 1910 and is best known for his work on streptomycin and other antibiotics,[20] work that earned him a Nobel Prize in 1952. Long before then, Waksman studied SOM formation. Although at first he accepted the humification hypothesis, he came out against it in a nearly 500-page tome published in 1936, in which he reviewed over 1300 studies, appearing in German, French, Russian, and Latin as well as English.[21] Such a review would be a lot of work to write today, even with the help of email and Google. The first sentence in the preface sets the tone and signals the book's scope:

> Although the importance of humus in agriculture was recognized by the early Greek and Roman philosophers, it is only during the last century and a half that attempts have been made to disclose the nature of this group of organic complexes, their formation and decomposition, and their role in plant nutrition.[22]

Later in the book, he states his objections to the way soil chemists had been studying soils and argues that harsh acid-alkaline extractions don't yield an accurate picture of humus and its "dynamic" nature. Instead, Waksman says "it should be recognized that the primary agents in the formation and transformation of humus are the microorganisms."[23]

Soil scientists didn't come around to Waksman's view for another 50 years. In their essay about Waksman and the humus problem, Philippe C. Baveye and Michelle Wander suggest several reasons why.[24] For many years, soil science was dominated by soil chemists who had few exchanges with soil microbiologists like Waksman, and it also took over a half a century to develop the tools, such as electron microscopy, nuclear magnetic resonance, and X-ray computed tomography, that have been essential in revealing SOM's true nature. Waksman's contributions have been overlooked, perhaps because he left the field in the 1930s and moved on to marine microbiology and his Nobel Prize–winning studies on antibiotics. In any event, his views have been vindicated today.

Soil scientists now think microbes and plants together are responsible for making SOM, without much input from the traditional humification pathway. Plants are still the source of organic carbon, but not just the plant parts from aboveground. Rather than just starting as detritus from leaves or branches, much of the carbon ending in soil is released by roots. This "belowground primary production" is about half of total production by plants.[25] Some organic chemicals released by roots are preserved, but not as large, complex macromolecules resulting from humification. Rather, the preserved plant chemicals are small and simple, which survive by adsorbing onto mineral surfaces or by just avoiding contact with microbes.[26] (Although each gram of soil has about a billion bacteria and dozens of meters of fungal hyphae, these microbes occupy only a small fraction of space within soil, less than a percent.) A simple sugar or a small protein without any protection would be quickly degraded by microbes, but it can survive for years tucked away in a tight soil crevice or stuck to a mineral, impossible for a microbe and its enzymes to reach. So these chemicals are old but not intrinsically refractory, and they would be easily degraded if they lost their mineral protection and came in contact with a bacterium or fungus.

Another major source of the organic chemicals in soils is bacteria and fungi. These microbes take up plant organic chemicals to grow and make more microbial biomass and cells. As they die, their cellular remains are released into the surrounding environment where they contribute to SOM. This mechanism of SOM formation has been supported by several lines of evidence. Soil microbial ecologists have labeled *E. coli*, a bacterial species common in the mammalian intestinal tract, with the stable (nonradioactive) carbon isotope of carbon, ^{13}C, by growing it on ^{13}C-labeled sugars. They then added the ^{13}C-labeled bacteria to soils and followed the ^{13}C as the bacteria died (they are unable to grow in soils).[27] After over 200 days, only a small fraction of the ^{13}C remained in the added bacteria while about half moved into nonliving SOM. As further evidence that bacteria contribute to SOM, soil microbial ecologists using electron microscopy saw fragments of bacterial membranes in the soil with the added ^{13}C-labeled bacteria. Other studies have found that soil matter contains amino sugars and other organic chemicals known to be made by bacteria and fungi. Like plant-produced chemicals, microbe-derived chemicals can be protected from degradation by absorbing onto mineral surfaces.

Chemicals like glucosamine and muramic acid known to be from microbes can be used to estimate how much SOM is from fungi and bacteria.[28] This dead microbial material of SOM is called necromass, analogous to biomass. Even though living microbial biomass is miniscule compared to SOM, microbial necromass is large. Overall, it accounts for roughly half of SOM, with the percentage varying with location, the type of soil, and the abundance of fungi versus bacteria. A study published in 2021 of over 900 sites mainly in North America, Europe, and China found that necromass made up 35 percent of SOM in the upper 20 centimeters of forest soils while it was over 50 percent in cropland topsoil. Relative necromass levels increased with soil depth in grasslands and forests but decreased in croplands, and they were higher in cooler soils or more acidic soils. In the 2021 study, fungal biomass was greater than bacterial biomass, so fungi contributed more to microbial necromass than did bacteria. Other microbial ecologists also think more necromass comes from fungi than from bacteria.[29] Whether fungi or bacteria, microbes explain a large fraction of SOM.

The way in which microbes die probably shapes the amount and chemical composition of necromass in soils (Fig 3.3).[30] A predator like a protozoan or nematode digests a bacterium or fungus mostly to carbon dioxide, leaving behind only partially degraded organic chemicals that could become necromass. By contrast, the organic chemicals left over from a bacterium killed by a virus may be largely intact. The other ways by which microbes die also probably leave their mark on necromass in soils. Death can come

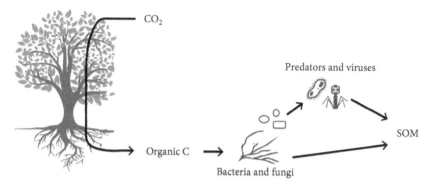

Figure 3.3 Formation of soil organic matter (SOM). Plant roots release organic carbon that can become SOM (not depicted here) or can be taken up by bacteria and fungi. These microbes then release compounds that become part of SOM or their cellular organic chemicals become SOM after they are eaten by predators. The cellular remains of microbes killed by viruses can also contribute to SOM.

to microbes that run out of nutrients, water, or time; microbes can undergo senescence and die of old age. Although soil microbial ecologists don't know which mortality factor claims the most microbes, it's clear that necromass is connected to the entire soil biota and to ecological interactions among many diverse organisms and viruses. There is much more to SOM than chemistry. Waksman would approve.

Regardless of how microbes meet their demise, microbial necromass in soils is a large stock in the global carbon cycle. Globally, necromass accounts for half of the carbon stored in soils and permafrost, over 1500 Pg C. That is much more carbon than what's in terrestrial vegetation or in the atmosphere as carbon dioxide. Not only are microbes responsible for some of the largest fluxes of carbon on the planet, but they are also behind one of the largest stocks as well. The smallest organisms are intimately connected to some of the biggest parts of the terrestrial carbon cycle.

A Changing Terrestrial Carbon Cycle

The global carbon cycle given in the first Intergovernmental Panel on Climate Change (IPCC) assessment[31] published in 1990 is similar in many respects to the most recent version in IPCC AR6. The diagram has the same boxes for the stocks and arrows for the fluxes, and their relative sizes are roughly the same.

But several numbers have changed substantially as the science has improved and more data have been gathered. Also, some of the numbers have changed because today's cycle is not the same one that operated over three decades ago. One of the more obvious changes is the amount of carbon dioxide in the atmosphere. It was 750 Pg C in 1990, rising to 871 Pg C in 2019, and the rate at which carbon dioxide increases each year went from 3 Pg C to 5 Pg C. Emissions continue to grow at a faster and faster rate. The unceasing rise in this greenhouse gas and others is why the planet is now threatened by climate change. Other changes in the carbon cycle since 1990 are a mix of good news and bad news.

The good news is on land. Terrestrial primary production and thus carbon dioxide consumption by land plants have been increasing for decades, according to several lines of evidence.[32] Based on eddy flux tower data from Europe and the United States, gross terrestrial primary production has been increasing by a percent each year since 1995.[33] Another study found a much earlier start date. Geochemists looked at Antarctic ice for bubbles that have trapped gases representative of the global atmosphere when the ice formed: going down deeper into the ice means going back further in time. One gas, carbonyl sulfide, is consumed by land plants like carbon dioxide and can be used to estimate primary production in the past. A study of carbonyl sulfide published in 2017 found that terrestrial primary production increased by 31 percent during the twentieth century.[34] That's good news because any carbon dioxide taken up by plants is carbon dioxide not in the atmosphere.

Primary production on land may be stimulated by our widespread use of nitrogen fertilizers, but climate scientists think production is also being enhanced by higher amounts of another type of fertilizer: atmospheric carbon dioxide. A recent study estimated that rising atmospheric carbon dioxide explains 44 percent of higher gross primary production while warming temperatures account for another 28 percent.[35] Both speed up a key enzyme in photosynthesis, ribulose 1,5-bisphosphate carboxylase/oxygenase, better known as Rubisco, the enzyme that fixes carbon dioxide into an organic chemical. Notoriously inefficient, Rubisco operates best when carbon dioxide levels are high. Higher levels also mean stomatal pores don't have to be wide open for plants to get enough carbon dioxide, so plants lose less water and grow faster. The fertilization effect is well-known to greenhouse operators who purposely raise carbon dioxide levels to boost plant yields, and it has been demonstrated more rigorously in greenhouse experiments

conducted by climate-change scientists. To test whether carbon dioxide fertilization occurs outside of greenhouses, plant ecologists have done free air
carbon dioxide enrichment (FACE) experiments in which an array of pipes
released pure carbon dioxide or carbon dioxide–enriched air along the pe-
rimeter of wide-open vegetated plots 8 to 30 meters in diameter,[36] and
then the response of plants was followed. Over 180 studies at 14 sites in five
continents demonstrated that higher carbon dioxide stimulates plant growth
on land.

The big unknown is the future. How long the carbon dioxide fertilization
effect will continue is unclear. It may depend on microbes.

One possibility is that carbon dioxide fertilization will diminish over
time as plant growth becomes more and more limited by nitrogen or phosphorus nutrients. That's one explanation for why the fertilization effect
lasted only the first few years at the FACE experiment conducted at the Oak
Ridge National Laboratory in Tennessee,[37] yet high carbon dioxide levels
continued to enhance growth of plants at other FACE sites. The difference
appears to be due to the severity of nitrogen limitation and the type of symbiotic fungi associated with plant roots, according to a review of data from 83
FACE experiments conducted in the United States, Europe, China, Australia,
and New Zealand. Growth of plants with ectomycorrhizal fungi continued to
be enhanced by elevated carbon dioxide levels regardless of nitrogen nutrient
levels, whereas the growth of plants with arbuscular mycorrhizal fungi was
slowed by low nitrogen levels and did not respond to high carbon dioxide.
Ectomycorrhizal fungi may be especially beneficial to their plant host when
nitrogen nutrients are low because they have enzymes to break down organic
nitrogen chemicals that arbuscular mycorrhiza cannot touch. This difference
in the metabolic capabilities of mycorrhizal fungi may explain why trees with
ectomycorrhizae respond more to high carbon dioxide levels than do grasses
with arbuscular mycorrhizae. Not all ectomycorrhizal fungi may be equally
helpful to their hosts. A recent study in Europe found some trees appeared to
grow faster with specific types of ectomycorrhizal fungi.[38] In short, mycorrhizal fungi may be key in determining how long carbon dioxide fertilization
and enhanced terrestrial primary production will continue in the upcoming
decades.

Primary production by terrestrial plants is only half of the story about the
future of carbon stored on land. The other half is respiration, in particular
respiration by soil microbes, and the news here is not so good. The response
of these microbes to climate change appears to be working against carbon

dioxide fertilization and its stimulation of land plant growth and carbon dioxide consumption.

Soil scientists had long expected that global respiration would increase over the years, but it wasn't until 2010 that the change could be measured. By that year, Ben Bond-Lamberty and Allison Thomson were able to compile four decades of results from chamber experiments,[39] the approach described in Chapter 2 used by soil microbial ecologists to measure soil respiration. Bond-Lamberty and Thomson found soil respiration increased significantly from 1989 to 2008 by about 0.1 Pg C each year. The rise in soil respiration was confirmed with newer data analyzed in a study published in 2021, although the story is complicated.[40] The main component of soil respiration, respiration by heterotrophic microbes, has increased over time, while the other component, respiration by roots, has not, according to the 2021 study. A change in soil respiration has serious consequences for the terrestrial carbon sink, depending on why soil microbes are respiring more.

The main explanation is global warming.[41] As temperatures have increased in most regions around the planet, so too has soil respiration, according to over 10,000 observations of respiration taken from 1961 to 2017 worldwide. A climate scientist would say there is "positive feedback" between soil microbes and temperature; higher temperature means more respiration and thus more atmospheric carbon dioxide, causing higher temperatures, and so on. In this case, positive feedback has quite negative consequences for the planet. The stimulation of respiration by global warming was expected because it was well-known that chemical reactions within microbes run faster as temperatures increase. Based on the correlation between temperature and respiration, models suggest that a warming of 2 degrees Celsius would cause soil organisms to release 10 Pg C of carbon dioxide more than normal, about the same amount of carbon dioxide emitted from fossil fuels each year.

The temperature effect on soil respiration also has been explored in experiments, some lasting for many years. Respiration has been followed in soils warmed by a buried steel-mesh heating pad or by warming the air above soils with giant heat lamps, to name two methods among several for heating soils. Respiration by warmed soils was then compared to controls without artificial warming.[42] In 50 of these experiments done around the world where the average temperature ranged from –18.1 to 19.6 degrees Celsius, a 2-degree warming stimulated soil respiration by 12 percent. That degree of stimulation is only slightly less than the 17 percent implied by the global correlation between soil respiration and temperature. Diving into individual experiments revealed some interesting exceptions and complications. In many experiments, soil respiration declined to the prewarming rate even when the temperature remained high, perhaps because microbes ran out of easily degradable organic carbon or another nutrient, or they became adapted to the warmer temperature. Some evidence arguing against the adaption hypothesis comes from an analysis that separated the response of soil microbes from that of plant roots. In those experiments, heterotrophic microbes usually continued to respire at higher rates in warmer soils over the years, whereas respiration by roots didn't change, suggesting that plants adapted to the higher temperature, but microbes didn't. That observation fits with other work indicating that heterotrophic microbes are more sensitive to temperature than are plants.

So global warming has stimulated respiration by bacteria, fungi, and other heterotrophs in soils over the last few decades. This is bad news. As the world continues to warm up, the carbon dioxide released by soil microbes threatens to undo the carbon dioxide fertilization effect that is stimulating higher carbon dioxide uptake by land plants. The severity of the problem depends on where soil bacteria and fungi get the extra organic carbon needed to fuel the temperature-enhanced respiration. In their 2010 publication first documenting the rise in soil respiration, Bond-Lamberty and Thomson pointed out that the enhanced respiration could be fueled by higher inputs of organic carbon recently made by plants. Primary production enhanced by higher carbon dioxide levels could mean roots release more organic carbon, stimulating respiration by soil microbes.

Another possible source of the extra organic carbon would have more serious implications for the terrestrial carbon cycle and thinking about climate change. Respiration jacked up by higher temperatures could be at the expense of the huge amount of carbon that has been preserved in SOM for decades to centuries. In a study published in 2018, Bond-Lamberty and colleagues looked at more soil respiration data and found some disturbing trends.[43] They saw the rise in soil respiration as others had seen and found that heterotrophic soil respiration, which is mainly by soil saprotrophic fungi and bacteria, increased more than root respiration did. What was new was their observation that the ratio of heterotrophic soil respiration to terrestrial plant production had increased as well. The title of a study published in 2021, which came to the same conclusion using a different analysis, says it all: "Greening of the earth does not compensate for rising soil heterotrophic respiration under climate change."[44] Even as primary production increases and land plants grow, soil microbes are respiring more and more carbon in SOM. Other data support that conclusion. In their 11-year experiment, Lei Cheng and colleagues saw the amount of organic carbon in both topsoil and subsurface soil (20–25 centimeters) decreased with warming, and their ^{14}C data indicate warming caused a loss of as much as 36 percent from SOM components that were decades to millennia old.[45] These investigators argue that as the planet warms, soil microbes tap into the huge pool of organic carbon that had been preserved in SOM, decomposing it to carbon dioxide and releasing it to the atmosphere.

It's still a bit unclear how bad SOM degradation would be for the terrestrial carbon cycle. Bond-Lamberty and colleagues acknowledged that data from eddy-covariance towers, which measure production and respiration above plants and soils, indicate that plant production has increased more so than soil and aboveground respiration has over the last few years. Perhaps temperature-enhanced respiration by soil microbes isn't a big problem. More data and time will tell. What's clearer today is how temperature affects the degradation of another stock of organic carbon on land, second in size only to SOM: the carbon now frozen in permafrost.

Carbon Bombs in the North

Thousands of gigatons of carbon are now frozen in permafrost covering millions of square kilometers in subsurface layers tens to hundreds of meters thick in Siberia, Alaska, and northern Canada. The amount of carbon now in permafrost is roughly equal to the amount in soils, and each is more than the carbon dioxide in the atmosphere (Chapter 2). Carbon needs to be frozen only for two years to be counted as permafrost, yet most of this carbon is much older, having accumulated over thousands to millions of years. With global warming, the frozen organic carbon is now thawing, a carbon bomb that could be triggered by microbes. The region with most of the world's permafrost, the Arctic, has been warming much faster than the rest of the world, especially in the northernmost zones where the average temperature of permafrost has been increasing by about 1 degree Celsius per decade.[46] That's more than enough warming to thaw permafrost and to transform it from rock-like to vulnerable ground prone to collapse and degradation.

Fancy scientific equipment isn't necessary to see the consequences of thawing permafrost (Fig. 3.4). People living in the Arctic have had little

Figure 3.4 Collapse of a coastal bluff along Drew Point, Alaska, due to thawing permafrost. Photograph by Benjamin M. Jones, University of Alaska Fairbanks.

choice but to build on permafrost because alternatives were too expensive, and permafrost seemed as permanent and secure as bedrock. When permafrost thaws, however, building walls crack, roads buckle, and cliffs slide into the ocean. In Nome, Alaska, a bridge lists ominously from uneven settling, and the runway at the small town's airport, constructed in a hurry during World War II, is sinking.[47] Thawing permafrost threatens not only over 500 kilometers of the Trans-Alaska Pipeline but also 1590 kilometers of the Eastern Siberia–Pacific Ocean oil pipeline and 1260 kilometers of gas pipelines in northwestern Siberia.[48] It has already caused a diesel spill near Norilsk City, Russia, after a storage tank built on permafrost failed.[49] Away from the Arctic, permafrost is found in high mountainous regions such as the Qinghai–Tibet Plateau, known in India as the Himalayan Plateau. Repairing the damage to the Qinghai–Tibet Highway caused by thawing permafrost cost about 4.5 billion Chinese yuan ($0.7 billion) between 1991 and 2011, roughly six times the construction price.

Thawing permafrost is also disrupting the global carbon cycle and potentially the planet's climate. Once preserved in a frozen state, thawed plant detritus and SOM now can be degraded by bacteria and fungi. Degradation can commence immediately as temperatures rise because bacteria, known to survive over millennia frozen in glaciers and in Antarctica,[50] become active when warming temperatures turn ice into liquid water. In addition to carbon dioxide, microbes also can release two other potent greenhouse gases, methane and nitrous oxide, in ground saturated with meltwater where oxygen is low or absent. But carbon dioxide release is disruptive enough. According to models compiled in the IPCC AR6, thawing permafrost will release as much as 41 Pg C as carbon dioxide for each degree that temperatures warm over the coming decades. One study estimated that 120 Pg C would be released by 2100, according to one scenario of a future climate envisioned by the IPCC.[51] The release could be enough to upset the balance of the carbon cycle on land, turning the Arctic from being a sink for carbon dioxide to a source. In a 15-year study of tundra west of Healy, Alaska, Ted Schuur and colleagues noted that gross primary production had increased over the years, but respiration increased more so, such that the Healy tundra has become a net source of carbon dioxide.[52] The switch occurred in the early 1990s when permafrost under the tundra started to warm and thaw. The Healy ecosystem used to help combat climate change, but as its permafrost thawed, it has been contributing to the problem. Climate scientists fear a similar fate for other Arctic regions.

Another, related carbon bomb is primed to go off, or at least smolder in lower latitudes, as well as in the Arctic. The bomb is peat, a type of soil rich in partially degraded vegetation that builds up at the bottoms of bogs and moors.[53] Like permafrost soils in the Arctic, peat is primarily found in northern Europe, Alaska, Canada, and Russia, with some in Southeast Asia, accounting for 25 percent of all soil carbon. Peat forms when plant detritus accumulates in waterlogged, acidic soils where the lack of oxygen prevents efficient decomposition. Peat contains so much carbon that its formation after the last ice age may have helped to prevent a rise in atmospheric carbon dioxide and warmer global temperatures.[54] When left undisturbed, today's peatlands take up carbon dioxide and are a net sink for greenhouse gases. But global warming may change peat into a net source of carbon dioxide and other greenhouse gases. Although peat in northern latitudes is especially susceptible to global warming, peatlands everywhere become vulnerable to burning when they dry out. Signs of the bomb going off are already apparent. Peat fires in Indonesia in 1997 and 1998 released about 0.95 Pg C, equivalent to about 15 percent of fossil fuel emissions during those years.[55] With less moisture and thus more oxygen, decomposition of peat carbon is faster, releasing more carbon dioxide. Estimates of how much carbon dioxide now being released by drained peatlands range from 0.31 to 3.4 Pg C of carbon dioxide per year, or 3 to over 30 percent of all anthropogenic carbon dioxide emissions. The worry is that even more carbon dioxide will be released if peatlands continue to be drained.

For now, however, for the terrestrial carbon cycle, the good news outweighs the bad. Despite the carbon bombs, terrestrial habitats are likely to continue to take up carbon dioxide from the atmosphere and slow the buildup of that important greenhouse gas. The stimulation of terrestrial primary production and carbon dioxide uptake by higher carbon dioxide levels is exceeding rising respiration and carbon dioxide release by soil microbes caused by global warming. The net effect is that land is still a substantial sink for carbon dioxide, even with thawing permafrost and draining peatlands. IPCC AR6 concluded with "high confidence" that the land sink for carbon dioxide continued to grow over the last six decades, implying that a large fraction of carbon dioxide released by human activity will continue to be taken up by land plants and stored in soils in the near future. As has been the case for several decades, uptake and storage by terrestrial habitats will slow the rise in atmospheric carbon dioxide and lessen climate change. The wild card is the carbon bomb in the north. Climate scientists working on IPCC AR6 have only "medium confidence" in their predictions about the response of permafrost carbon to global warming. Located often deep underground in remote regions with harsh climates, permafrost is difficult to study. Peatlands are another carbon bomb. SOM and microbes may be easier to study in lower latitude regions, but even there the effect of climate change is not completely understood. What is clear is that the size of the terrestrial carbon sink depends in large part on microbes and their response to climate change.

4

Carbon Pumps in the Oceans

The world as we know it wouldn't exist if water weren't so unusual and ice-nine were real. In Kurt Vonnegut's dystopian science-fiction novel *Cat's Cradle*, ice-nine is an alternative form of water that is a solid at room temperature, and when added to normal water, turns it into ice. Near the end of Vonnegut's book, a plane carrying ice-nine crashes, and the world's seas, rivers, and lakes freeze solid, dooming civilization and the planet. Vonnegut got the ice-nine idea from his brother Bernard, an atmospheric scientist who had discovered that seeding clouds with silver nitrate could produce rain.[1] (There really are different forms of ice, including ice-nine, along with ice-one, or everyday ice, through ice-twenty.[2]) Although Earth can do without Vonnegut's ice-nine, it can't do without the normal version of water with all its unusual properties. Unlike other liquids, water is most dense and heaviest at 4 degrees Celsius, so ice is light and floats, providing a habitat for polar bears and penguins and a white surface that reflects sunlight and heat away from the planet. If the cohesion between water molecules was not so unusually tight, the temperatures at which ice melts and water boils would be lower, and there would be much less liquid water on Earth. Depending on how they are counted, there are as many as 75 anomalous properties of water.

Several of these properties, along with the fact that seawater covers so much of the planet, explain why the oceans are so important in thinking about climate change. One property is that water can take up a lot of heat. Because of water's high heat capacity, the oceans have absorbed about 90 percent of the heat trapped by the atmosphere over the last century, which has raised the amount of heat in the oceans by about 10-fold.[3] Not only has the average temperature of the surface ocean crept up by almost 1 degree Celsius worldwide over the last 100 years,[4] marine heatwaves have become more common. Globally, over the last 40 years, the frequency of extreme heat episodes in the oceans has doubled, and they have lasted over 80 percent longer.[5] Long periods, some stretching for hundreds of days in the eastern tropical Pacific for example, of abnormally warm surface waters have devastated coral reefs and seagrass meadows, caused massive die-offs of marine mammals and birds, and prolonged harmful algal blooms in coastal waters.[6] This bad news for the oceans means, however, there is a bit of good news for us on land. The heat being absorbed by the oceans means less is affecting terrestrial habitats. Global warming on land would be even worse if not for the oceans.

Another anomalous property of water is that it's an excellent solvent, enabling the oceans to suck up a lot of carbon dioxide from the atmosphere, almost as much as the sinks on land do. But they do it by a very different mechanism. While the terrestrial sink starts with land plants taking up carbon dioxide from the atmosphere during photosynthesis, the marine sink starts with the diffusion of atmospheric carbon dioxide into surface waters. The now dissolved carbon dioxide (CO_2) then reacts with water to form bicarbonate (HCO_3^-) and carbonate (CO_3^{2-}); the sum of all three carbon chemicals is known as dissolved inorganic carbon (DIC). The ocean would store very little carbon if it were filled with a liquid other than water and if carbon dioxide were like gases that don't react with water. For the ocean to be a sink, however, the inorganic carbon must be moved into deep waters out of contact with the atmosphere. Carbon in surface layer easily moves back into the atmosphere. How carbon gets to deep waters varies among the different kinds of carbon pumps operating in the ocean. These mechanisms are called "pumps" because they move carbon from the surface with relatively low carbon dioxide concentrations to deep waters where concentrations are much higher. Of the three pumps known to oceanographers (solubility pump, biological pump, and microbial carbon pump), we'll start with the one that doesn't involve microbes.

The solubility pump depends solely on chemistry and physics to transport carbon away from the atmosphere. After diffusion of carbon dioxide into surface water and its chemical transformation to DIC, physics takes over to move the carbon to greater depths. That happens during deep water formation. In northern lobes of the North Atlantic Ocean and off Antarctica, surface water loses heat and evaporates, leaving behind cooler, saltier, and thus denser water that sinks to the bottom. The newly formed deep water then flows south or north to lower latitudes as part of thermohaline circulation, informally known as the conveyor belt.[7] As surface water sinks, it carries everything with it to the bottom, including the carbon dioxide it had picked up from the atmosphere. Earth is fortunate that deep water forms at high latitudes where it's cold. Cold water holds more carbon dioxide and other gases than warm water can, explaining why soda, beer, and champagne go flat and the fizz disappears when those carbonated beverages warm up. Because deep water forms in high latitudes where surface water is frigid and carbon dioxide solubility is highest, the solubility pump can take a lot of carbon dioxide out of the atmosphere and tuck it away in the deep ocean for centuries.

The solubility pump is a reason why deep waters have higher concentrations of DIC than surface waters do, but it's not the only reason. It's not even the main reason, as it explains only about 10 percent of the difference between surface and deep DIC concentrations.[8] The remaining 90 percent is due to the biological pump (Fig. 4.1). Without it, atmospheric carbon dioxide would be 150 to 250 parts per million higher,[9] over 30 to 60 percent higher than the current level. Of the many reports about the strength of the biological pump, here I will rely on a recent study by Michael Nowicki and colleagues that combined satellite and oceanographic data and a numerical model to produce numbers for the entire biological pump and its components.[10] Nowicki's study says the biological pump exports about 10 petagrams of carbon (Pg C) each year, roughly the same amount of carbon dioxide put into the atmosphere every year by human activity.

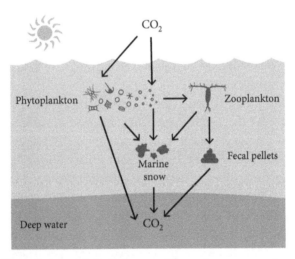

Figure 4.1 The soft tissue biological pump. The main components are the sinking of large phytoplankton cells, zooplankton fecal pellets, and aggregates ("marine snow") composed of detritus from phytoplankton, zooplankton, and bacteria. When organic matter reaches deep waters, it is degraded by microbes to carbon dioxide (CO_2). Not shown here is the carbon exported by the migration of large zooplankton to deep waters.

This being the oceans, much of the biological pump is microbial. The pump starts in sunlit surface waters where cyanobacteria and eukaryotic microalgae take up dissolved carbon dioxide and make organic cellular components. Some of this organic matter then sinks to deep waters. How deep it sinks and how long it stays in deep waters help to determine the efficiency of the biological pump. How the organic matter sinks varies, but the end is the same for all: carbon dioxide is exported to deep waters, sequestering it away from the atmosphere for decades to centuries.

The biological pump can be divided into two parts. In the biggest, the "soft tissue" biological pump, the exported carbon is organic matter made by phytoplankton. It makes up 70 percent of the sum of both the biological pump and the solubility pump. Chapter 6 will go over another part of the biological pump, called the hard tissue pump or the carbonate pump, in which the exported carbon is calcium carbonate, the cement-like material made by some algae (coccolithophores) and a few other microbes.

Marine Snow and Zooplankton

The biological pump solved a problem that stumped scientists in the nineteenth century. In 1841 Edward Forbes, the naturalist aboard the *H.M.S. Beacon*, dredged the bottom of the Aegean Sea and found that the diversity, abundance, and size of marine biota decreased with water depth.[11] Extrapolating to deeper waters where he didn't sample, Forbes argued that life stops entirely at 300 fathoms (about 550 meters). Marine organisms couldn't exist, according to Forbes's azoic theory, in waters so cold, under such great pressure, and without food from light-dependent algal growth at the surface. The *H.M.S. Challenger* expedition conducted from 1872 to 1876 laid the azoic theory to rest when it found marine life thriving in some of the deepest parts of the ocean. After writing a dry account about echinoderms for the *Challenger* report, Alexander Agassiz published a book about deep-sea life, based on his work on three cruises abroad the U.S. Coast and Geodetic Survey Steamer *Blake*, which explored the Gulf of Mexico and the U.S. East Coast in the late 1870s.[12] He described how animals at the bottom of the ocean were fed by dead matter from the surface sinking to deep waters, what he supposedly called a "rain of detritus."[13]

By the time it was actually seen 50 years later, Agassiz's rain had turned into snow, metaphorically speaking. The snow metaphor was first used

in Rachel Carson's book *The Sea Around Us*, published before her most famous book, *Silent Spring*, which launched the modern environmental movement in 1962. Carson had an undergraduate degree in biology (unusual for
a woman in 1929) and a job at the U.S. Bureau of Fisheries, yet she wanted
first-hand experience of subsurface life to help her write *The Sea Around Us*.
She arranged to meet William Beebe, who was famous at the time for trips
on his bathysphere into deep waters off Bermuda and who eventually would
become one of Carson's influential and ardent supporters.[14] But Carson
didn't get to Bermuda and had to settle for the Florida Keys. Her "dive" there
consisted of descending into the water while holding onto the boat's ladder,
dressed in work pants and a long-sleeved shirt over her swimsuit, topped off
by an 84-pound diving helmet. Her first and only venture below the sea surface was marred by bad weather, and she could see little, saying it was like
"London fog."[15] But it was transformative. She made up for her limited direct
experience with deep dives into the literature, conversations with the experts
like Beebe, and a vivid imagination. In *The Sea Around Us*, Carson described
life at the bottom:

> When I think of the floor of the deep sea, the single, overwhelming fact that
> possesses my imagination is the accumulation of sediments. I see always
> the steady remitting, downward drift of materials from above, flake upon
> flake, layer upon layer—a drift that has continued for hundreds of millions
> of years, that will go on as long as there are seas and continents. For the
> sediments are the material of the most stupendous "snowfall" the earth has
> ever seen.[16]

The first reports in the scientific literature about Carson's snowfall were
by Japanese oceanographers working in the late 1940s and early 1950s, although long before then Japanese fishermen were familiar with "nuta" that
sometimes coated their fishing nets and was taken as a sign of productive
waters.[17] In the late 1940s, Tokimi Tsujita of the Seikai Fisheries Research
Laboratory collected samples from surface waters off Tsushima Island, at
the southern end of the Japanese archipelago, and with the help of light and
electron microscopy, concluded that nuta was the "skeletal remains of ex-
plosively grown population of plankton and living plankton attached to the
suspended matter."[18] Working about the same time off Hokkaido, Noboru
Suzuki and Kenji Kato observed "numerous flake-like substances" through
the porthole of the *Kuroshio*,[19] a tiny submersible equipped with a small bird

to monitor air quality. In the opening paragraph of their 1953 publication, Suzuki and Kato wondered if the substances they saw were like Carson's "stupendous snowfall in the sea," and a couple paragraphs later, they propose the name "marine snow." It was a better, more informative name than the alternative, "leptopel," which had been proposed earlier by Scripps Institution of Oceanography scientists.

Suzuki and Kato's description of what makes up marine snow still holds today: "it is clear that the suspended materials are chiefly the aggregates of the remains of plankton, sinking in some stages of disintegration by marine bacteria." The aggregates are held together by mucus from gelatinous zooplankton or by polysaccharide-rich polymers released by phytoplankton and bacteria. As these aggregates sink, they pick up smaller particles and accrete. The particles swept into the aggregates can include other detrital particles, small phytoplankton, bacteria, and hard mineral parts of algal cell walls, such as calcium carbonate from coccolithophores and opal, a glass-like mineral found in diatoms. Acting as ballast, the minerals increase the density of the aggregate. As the marine snow aggregate grows and becomes heavier, it sinks faster and faster through the water column. Yet few escape being eaten by zooplankton or decomposed by bacteria to survive long enough to reach deep waters. Those that do make up the biological pump.

Like sinking aggregates, diatoms, coccolithophores, and a few other phytoplankton are big and dense enough to sink. Large phytoplankton are also inclined to form aggregates when they can reach high concentrations during phytoplankton blooms, the result of rapid growth in the nutrient-rich waters found along some coasts. Although small phytoplankton like single-cell cyanobacteria do not sink appreciably, they can be swept up into aggregates and contribute to the biological pump. Sinking aggregates and marine snow are important in some regions, such as coastal waters that host phytoplankton blooms, where they contribute as much as 40 percent of the biological pump, but globally they account for only about 15 percent.[20]

The biggest part of the biological pump is run by large zooplankton that feed at the surface and excrete fecal pellets which sink into deep waters. As with all parts of the biological pump, the carbon in fecal pellets starts as carbon dioxide that is fixed into organic matter by phytoplankton, which is then repackaged by zooplankton into particles large enough to sink. Only big zooplankton like copepods and krill, metazoans 0.2–20 millimeters in length, excrete fecal pellets large enough to sink, unlike the excrement from small zooplankton and protists that stays suspended where it is released.

Along with their size, the cigar-like shape of big zooplankton pellets, formed during passage through the animal's digestive tract, enhances its descent to deep waters. As with aggregates, most of the fecal pellets don't make it out of the surface waters where zooplankton feed. The pellets are degraded either by bacteria colonizing the pellets or by zooplankton that eat the sinking pellets, an example of coprophagy. Enough of the fecal matter, however, does survive the bacterial and zooplankton gauntlet, such that in some oceans, sinking pellets make up as much as 90 percent of the biological pump; overall, they account for 80 percent. Zooplankton also help the biological pump by migrating from the surface at night to deep waters during the day where they excrete organic matter or exhale carbon dioxide. This daily migration contributes another 10 percent of the pump.

Sinking fecal pellets and migrating zooplankton don't have many connections to microbes, except for the fact that everything starts with primary production by phytoplankton. Microbes are more obviously the main players in the final part of the biological pump to be discussed here: the export of dissolved organic carbon (DOC).

"Gatekeepers of the Biological Pump"

DOC in the ocean is one of the largest pools of carbon on the planet, about the same size as atmospheric carbon dioxide, only somewhat smaller than the pool of organic carbon locked in soils. (That similarity between oceanic DOC and soil organic carbon won't be the last we'll see.) As with the other parts of the biological pump, the carbon in DOC is from carbon dioxide fixed by phytoplankton into organic chemicals during photosynthesis. Some of those chemicals are released immediately by phytoplankton into the DOC pool while others are transferred up the food chain when phytoplankton are eaten by herbivorous protists and zooplankton. Those organisms also release DOC, as do viruses when they infect and split open, or lyse their victims. Every step along the food chain can produce DOC. Nearly all this new DOC, freshly released by organisms and viruses, what microbial ecologists call labile DOC, is easily degraded and is consumed within minutes to days mostly by heterotrophic bacteria. Because it is used so quickly, labile DOC is less than a percent of the total DOC pool.[21]

Even though labile DOC is a very small fraction of the total, the movement or flux of carbon through the DOC pool is large. Roughly half or more of primary production is routed from phytoplankton through labile DOC to heterotrophic bacteria and onward to their protist predators and the rest of the food chain.[22] This pathway is called the microbial loop (Fig. 4.2). It's a huge part of marine food chains. Although biological oceanographers knew about marine bacteria since at least the 1890s, they thought the carbon cycle and marine food chains were run by large phytoplankton and zooplankton.[23] Bacteria were thought to be on the periphery, their role confined to degrading dead detritus and recycling nutrients to support phytoplankton growth. Now we know that in many oceanic regions, most phytoplankton are tiny like *Prochlorococcus*, other cyanobacteria, and small photosynthesizing protists 5 to 20 microns in diameter. Some of these protists are also the main herbivores, leaving copepods and other big zooplankton to prey on the few large phytoplankton. What distinguishes the microbial loop from other marine food chains is not just the size of the organisms but the central role of DOC. Rather than a particulate form of primary production that can be eaten by protists and zooplankton, a large fraction of primary production is routed through a dissolved form that is consumed mostly by heterotrophic bacteria. (Unlike in soils, fungi are outcompeted by bacteria and aren't very abundant in the oceanic water column, so they don't account for much DOC degradation.)

The microbial loop is a weak starting point for the biological pump and carbon sequestration. Any carbon routed through the loop must survive several steps before reaching a sinking aggregate or a zooplankton big enough to

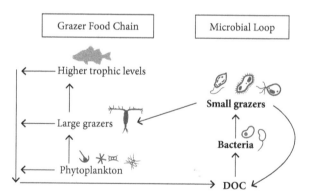

Figure 4.2 The grazer food chain and the microbial loop. DOC is dissolved organic carbon. Large grazers are zooplankton like copepods. Protists are examples of small grazers.

excrete a fecal pellet or to migrate to deep waters. Each step has a cost: carbon dioxide released during respiration. Heterotrophic bacteria alone respire a lot; 70 percent or more of the DOC they assimilate is broken down to carbon dioxide.[24] Any primary production respired back to carbon dioxide by bacteria and other microbial loop microbes takes away carbon that could have been exported by the biological pump. Viruses don't help. A bacterium or other microbe killed by a virus is removed from the food chain that eventually could produce a fecal pellet or an aggregate big enough to sink to deep waters. It doesn't help the biological pump much when the organic remains of the virus-lysed microbe are returned to the DOC pool.[25] Once back in a dissolved state, a carbon molecule is more likely to be lost as carbon dioxide during respiration than to contribute to the pump. Hugh Ducklow, a microbial oceanographer who has worked on connections between the microbial loop and the biological pump for most of his career, has called bacteria "the gatekeepers of the biological pump."[26] Bacteria keep the gate mostly closed when they degrade organic carbon and respire it to carbon dioxide, not allowing much carbon to enter the grazer food chain and the biological pump.

A carbon molecule can slip by the bacterial gatekeeper if it makes its way to forms of DOC that are not readily used by bacteria or other microbes. Most DOC in the oceans is refractory and resists degradation by microbes for months to tens of millennia.[27] If it weren't refractory, the DOC pool would be much smaller than it is now, and atmospheric carbon dioxide would be nearly double its current level (Chapter 2). Chemical oceanographers have deduced the age of DOC and how long it resists degradation by looking at data from the ^{14}C-dating method (see Box 3.1). According to ^{14}C data, the age of the total DOC pool in the deep ocean is about 6000 years.[28] That is the average. Some chemicals making up deep ocean DOC are much younger, while others are much older, as much as 30,000 years old.[29] The lifetime of DOC in the ocean is similar to how long organic matter lasts in soils (Chapter 3). It's amazing that any organic chemical could last so long in seawater surrounded by starving bacteria without much protection by mineral surfaces, unlike in soils.

Between the two extremes set by labile DOC and refractory DOC is semi-labile DOC (Fig. 4.3). In the surface layer, DOC can build up over the summer, reaching up to about 1.0 milligrams per liter in productive coastal waters. The surface-layer DOC that accumulates over the year is called semi-labile DOC, because its susceptibility to degradation is between labile

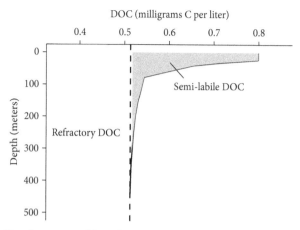

Figure 4.3 Two fractions of dissolved organic carbon (DOC) in the oceans as defined by how DOC varies with depth. The concentration of refractory DOC at the surface is the same as in deep waters (the dashed lane). Semi-labile DOC (the shaded region) is what builds up during the summer above the seasonal thermocline, which is a density barrier that prevents mixing of the surface layer with deep waters. Concentrations of labile DOC are too low to show up here. The data used here are from the Ross Sea, Antarctica, given in Dennis Hansell's 2013 review.

DOC and refractory DOC. By surviving for months, semi-labile DOC is more refractory than labile DOC, which is immediately consumed by heterotrophic bacteria, but it's more labile and disappears faster than 6000-year-old refractory DOC does. Nearly all DOC in deep waters is refractory, so the question is about the amount of semi-labile DOC in the surface layer. Key to answering that question is the observation that refractory DOC concentrations must not change much with depth because refractory DOC is older than the residence time of water in the global ocean (6000 years or more versus about 3000 years). (See Box 4.1 for more about residence time.) Refractory DOC is so old that it travels through the world ocean from top to bottom a couple times or more, thoroughly mixing up and down the water column, before it finally disappears. So the concentration of refractory DOC in surface water is the same as what it is in deep water. If that is the case, then the concentration of semi-labile DOC is the difference between the concentration of total DOC in the surface layer and the DOC concentration in deep water—the refractory DOC concentration. (There is too little of labile DOC to matter in these calculations.) We need to know about this because the

---

Box 4.1 How Old Is Seawater?

Residence time combines two important concepts in biogeochemistry: pool size and the flux of material going into and out of that pool. Here, residence time is a measure of how long an average parcel of water remains in the ocean after traveling around the world from the surface to the ocean floor. Water is added to the ocean as precipitation, groundwater, or runoff from land and returns to the atmosphere via evaporation at roughly the same rate (4.3×10^5 cubic kilometers each year). Dividing that rate into the volume of the ocean (1.3×10^9 cubic kilometers) yields a residence time of about 3000 years.

---

semi-labile fraction contributes the most of all DOC components to the biological pump.[30]

But to be part of the biological pump, semi-labile DOC needs to get to deep waters. Like the export of DIC by the solubility pump, export of DOC and its contribution to the biological pump depends on physics. Surface water with its semi-labile DOC needs to sink to deeper depths or otherwise degradation would return carbon dioxide back to the atmosphere. The discussion of the solubility pump focused on deep water formation and the sinking of surface water to the abyss in a few high-latitude regions. DOC is exported in those high-latitude oceans but in other oceans as well. In other regions, winds can push surface water down deeper, although not all the way to the bottom, carrying DOC to below the surface layer. This wind-driven push is a big reason why DOC export is 30 to 50 percent of the biological pump in subtropical seas.[31] For the entire global ocean, DOC export is 1.9 Pg C each year, accounting for about 20 percent of the total biological pump. However, except in regions where DOC is carried to the abyss during deep-water formation, DOC is not pumped deep enough to be out of contact with the surface for long. The carbon in exported DOC is sequestered only for about 50 years on average for the global ocean. By contrast, the carbon exported by zooplankton fecal pellets and migration goes deeper and is out of contact with the surface layer for as long as 150 years.

Refractory DOC, the largest fraction of DOC, doesn't contribute much to the biological pump because the concentration of this very old carbon is the same from the surface to the bottom of the water column. Export is only possible if concentrations are higher in the surface layer than below it. The mixing of surface water with deep water, both with about 0.5 milligrams of refractory carbon per liter, doesn't change anything. Although refractory DOC doesn't contribute to the biological pump, it still sequesters a lot of carbon away from the atmosphere by a different mechanism. This sequestration mechanism is called the microbial carbon pump.[32]

The Microbial Carbon Pump

This pump differs from the others. The solubility pump and the biological pump sequester carbon away from the atmosphere by moving it to deep waters. The microbial carbon pump doesn't move carbon anywhere except into chemicals that can't be transformed easily to carbon dioxide. Although refractory DOC can be ignored when discussing the biological pump, it's a major part of the microbial carbon pump. Refractory DOC houses a lot of carbon that otherwise could be in the atmosphere as carbon dioxide. Human activity has added a huge amount of the gas, 280 Pg C, to the atmosphere over the last 200 years, but that is smaller than the 700 Pg C now locked in refractory DOC.[33] Much of refractory DOC exists because of the microbial carbon pump. The term earns the "microbial" part not because bacteria and other microbes are so great at degrading refractory DOC. They aren't. What little degradation may not be by microbes but by intense heat when refractory DOC is sucked below the seafloor near hydrothermal vents or by bright sunlight when at the surface on one of DOC's trips around the ocean.[34] No, "microbial" highlights a possible mechanism for making refractory DOC.

To understand refractory DOC, at first marine organic geochemists borrowed some of the same methods and terminology used by soil chemists (see Chapter 3). (Selman Waksman, that Nobel Prize–winning soil microbiologist who eventually came to question soil humics, worked on "marine humus" in coastal waters of Cape Cod, Massachusetts, and the Gulf of Maine.[35]) Marine geochemists thought refractory DOC started as labile organic chemicals made by phytoplankton that were transformed by condensation and other chemical reactions into complex humic matter difficult for bacteria to decompose. As soil chemists moved away from the humification hypothesis, so too did marine organic geochemists when they found that most DOC, especially the refractory parts, was small, roughly 1000 Daltons, unlike the large molecules predicted by the humification hypothesis.[36] (A protein may be 100,000 Daltons while table sugar is 342 Daltons. Small chemicals like table sugar and amino acids are easily decomposed and are not refractory DOC.) Also, new chemical analyses found organic chemicals other than humus, such as carboxyl-rich alicyclic molecules, somewhat better known as CRAM.[37] The deduced structure of CRAM is like that of terpenoids, a diverse class of biochemicals that includes sterols and other components of membranes.

Bacteria appear to be a major source of these membrane components. Chemical analyses have turned up chemicals like muramic acid, lipopoly-saccharide, and D-amino acids that come from membranes or cell walls of bacteria. Muramic acid is found only in bacteria, while lipopolysaccharide is unique to Gram-negative bacteria and cyanobacteria,[38] the most abundant bacteria in the oceans. Whereas L-amino acids, the most common optical isomer of amino acids, make up protein in all organisms, D-amino acids are in cell walls of bacteria. Reminiscent of work by soil microbial ecologists, experiments conducted by marine microbial ecologists have shown that after taking up labile organic chemicals, bacteria release refractory DOC.[39] In one experiment, two labile chemicals, a sugar or an amino acid (glucose or glutamate, respectively), were fed to a marine bacterial community and the concentrations of the added chemicals and total DOC were followed over time. The sugar and amino acid were depleted within days, mostly respired by bacteria to carbon dioxide. DOC concentrations also dropped but not as much as expected from the decrease in the two labile chemicals, suggesting bacteria produced unknown organic matter resistant to degrada-tion. Although concentrations of this unknown, refractory DOC decreased during the second phase of the experiment, as much as 50 percent remained after 1.5 years when the experiment ended. A DOC component surviving for a year or two may not be truly refractory, capable of lasting for millennia, but it is still impressive that bacteria transformed the labile chemicals to dissolved matter that was over 300 times more resilient to degradation.[40] Based on chemicals like D-amino acids that trace organic matter from bacteria, Ron Benner and Gerhard Herndl estimated that bacteria make about 25 percent of total DOC and 50 percent of semi-labile and semi-refractory DOC.[41]

This is the microbial pump. Labile organics produced by phytoplankton are taken up by heterotrophic bacteria, which then release, probably during protist grazing and viral infections, dissolved organic chemicals that end up as semi-labile or refractory DOC.

Even if correct, as seems likely, the microbial pump does not explain why the bacteria-produced DOC resists degradation. Several hypotheses have been suggested. Some marine organic geochemists and microbial ecologists believe the intrinsic structure of organic chemicals makes them refractory or not. That is the case for black carbon, which is like charcoal produced by combustion of terrestrial plant biomass. Black carbon can be extremely old, having escaped degradation for more than 30,000 years. Other geochemists and microbial ecologists have argued that DOC is refractory not necessarily because of its structure but because individual DOC molecules are extremely rare or dilute and extremely diverse, so much so that bacteria cannot use them; this idea is called the dilution hypothesis.[42] Even though the DOC pool for the entire world ocean is large, it's mostly because the ocean is huge; in fact, the amount of DOC in a liter of seawater is low. A liter has only about 0.5 milligrams of carbon as refractory DOC. (A liter of Coke has 106,000 milligrams of sugar.) It is also true that there are many different types of DOC molecules, more than 600,000 according to high-resolution mass spectrometry. A few marine organic geochemists have combined these ideas and have argued that some DOC may be refractory because of both its chemical structure and its diluted state (the two hypotheses are not mutually exclusive), while also emphasizing that the refractory nature of a chemical depends on environmental conditions and the state of the microbial community. Oil, for example, is preserved for millions of years when locked in oxygen-free reservoirs but can be decomposed relatively quickly by petroleum-degrading bacteria when it leaks or is spilled into oxygen-rich waters.

To recap, microbes are the main agents that drive uptake and sequestration of carbon dioxide by oceanic biology. Both the biological pump and the microbial carbon pump start with the consumption of carbon dioxide by phytoplankton, which transform the dissolved gas into forms of organic carbon matter that prevent carbon from exchanging with the atmosphere. Some microbial activities diminish the biological pump while others strengthen it. Heterotrophic bacteria decompose the main organic matter making up the biological pump—fecal pellets, marine snow, and other aggregates—before it can escape the surface layer and reach deeper waters. But other microbes add to the biological pump through the production of sticky polymers that help aggregates to form and sink. Finally, the microbial carbon pump is all microbes. Formed by this pump, refractory DOC sequesters carbon dioxide away from the atmosphere as effectively or more so than if the carbon were stored in the deepest ocean. This refractory DOC is like the refractory organic carbon in soils. A large fraction of both pools can be traced back to microbes, and the size of each is huge, containing more carbon than what is in the atmosphere. If ExxonMobil or Shell found a way to refine refractory organic carbon and make gasoline or oil, burning it would disrupt Earth's climate as seriously as the combustion of fossil fuels is doing today.

The Biological Pump in the Past

Earth is absorbing some of the insults brought on by the climate change humans have caused. The heat sucked up by the ocean has lessened global warming on land, and carbon dioxide levels in the atmosphere would be increasing even more if land and the ocean together didn't take up about half of what's released by human activity.[43] Although land and the ocean consume nearly the same amount of the released carbon dioxide (29 percent versus 26 percent of total emissions, respectively), how they do it couldn't differ more. Plants are doing the consumption on land while in the ocean, it is the solubility pump, that physical–chemical mechanism which starts with the diffusion of carbon dioxide into seawater at the surface, followed by the sinking of surface seawater to deep depths. The strength of the land and ocean sinks for carbon dioxide has grown over the years but for different reasons. On land, rising atmospheric carbon dioxide levels stimulate terrestrial plants (with help from fungi and other microbes) to take up more of the greenhouse gas (Chapter 3) while in the ocean, the rising levels force more carbon dioxide to diffuse from the atmosphere into the surface ocean where it can be moved into deep waters. There is concern that the land and ocean sinks may weaken in the future if climate change continues along its current course. For now, at least, both sinks are going strong, to our benefit. The rise in atmospheric carbon dioxide would be even greater and climate change worse if not for the carbon dioxide taken up by plants on land and by the solubility pump in the ocean.

What about marine microbes and the biological pump? Since the biological pump is responsible for storing more carbon in the deep than what the solubility pump puts there, it seems that microbes should be a big part of the story. The short answer is the biological pump isn't doing anything with the carbon dioxide released by human activity. Why it's not is a bit unclear. Because cyanobacteria and eukaryotic microalgae carry out photosynthesis mostly like terrestrial plants, we would expect higher carbon dioxide levels would lead to higher rates of photosynthesis in the ocean, just like the carbon dioxide fertilization effect seen on land. Higher concentrations of carbon dioxide would alleviate the need for expensive "carbon-concentrating mechanisms" used by some phytoplankton to get more carbon dioxide for the carbon dioxide–fixing enzyme in photosynthesis, ribulose 1,5-bisphosphate carboxylase/oxygenase. The enzyme is just as inefficient in algae and cyanobacteria as in land plants. With more photosynthesis, phytoplankton would grow faster, rates of primary production would be higher, and the biological pump would be stronger. Some phytoplankton do grow faster with higher carbon dioxide levels, just like land plants, but more do not, or are even inhibited.[44] Perhaps phytoplankton are prevented from growing faster because of low nutrients, or maybe they already adapted to high carbon dioxide levels that can occur naturally in some oceanic regions. In any case, unlike the response of land plants, rising atmospheric carbon dioxide is not enhancing primary production nor the biological pump in the ocean.

Just the opposite. Climate change may have already caused oceanic primary production to decrease. Daniel Boyce and colleagues were among the first to argue that oceanic phytoplankton biomass has declined over the last century,[45] according to their analysis of data on chlorophyll, an index of phytoplankton biomass, and water clarity, another index of phytoplankton that has been measured since the nineteenth century, thanks to Father Pietro Angelo Secchi. When phytoplankton become more abundant, water turns more turbid and the depth at which a 20-centimeter white Secchi disk disappears from sight becomes shallower. Since Father Secchi invented his eponymous disk in 1865 to explore the clarity of the Mediterranean Sea for the Papal Navy, Secchi disks have been lowered into most of the world's oceans for over a hundred years.[46] Oceanographers need every year of data they can get if they are to distinguish a long-term trend potentially fueled by climate change from year-to-year variation in phytoplankton that may cycle up and down for reasons unconnected to the anthropogenic climate change we're focusing on here. Boyce and colleagues used the Secchi disk data and numbers from other sources to argue that oceanic primary production has decreased over the last 100 years. Other studies have found that changes depend on the region and the years examined. From 1998 through 2013, for example, chlorophyll decreased in gyres of the North Pacific, Indian Ocean, and the North and South Atlantic, while there was no significant trend in the South Pacific.[47] So, unlike land plants, if anything, oceanic phytoplankton are doing worse with climate change.

Along with looking at how climate change affects phytoplankton and primary production, we need to consider how the biological pump could change the planet's climate. As pointed out before, atmospheric carbon dioxide levels would be much higher if not for the biological pump, so variation in the pump's strength or efficiency may have big impacts. Even though there's not much evidence that the pump is changing today, it apparently did in the geological past. There is some evidence that changes in the microbial carbon pump contributed to the Paleocene-Eocene Thermal Maximum about 56 million years ago when the planet warmed by about 5 degrees Celsius.[48] But we know more about temperature swings connected to the biological pump during the last 650,000 years, as revealed by air bubbles and material trapped in Antarctic ice. The European Project for Ice Coring in Antarctica (EPICA) project recovered ice cores extending nearly 3300 meters below the surface and found evidence of eight ice ages, or glacial-interglacial cycles, over 650,000 years before the present.[49] Carbon dioxide, measured directly in ice bubbles, and temperature, deduced from hydrogen and oxygen isotopes in water, varied together over the eight cycles; when carbon dioxide levels were high, temperatures were high, and vice versa. Although a recent study using the triple-oxygen isotope method argued for the importance of terrestrial primary production for explaining the cycles,[50] other studies have pointed to the biological pump in the Southern Ocean as being key to understanding why ice ages have come and gone over the last 650,000 years.[51]

The Southern Ocean is key because its surface waters have persistently high concentrations of nitrate and phosphate—the major nutrients needed by all phytoplankton—and of silicate required by diatoms. What puzzled biological oceanographers for decades was that despite having enough of the major nutrients, chlorophyll and by implication phytoplankton biomass is relatively low in the Southern Ocean. It's the biggest ocean among a few others with high nutrients but low chlorophyll. With high nitrate, phosphate, and silicate, growth of phytoplankton should be fast, biomass high, and the biological pump strong, but they're not—not today, at least. The high nutrient–low chlorophyll puzzle was solved when John Martin and colleagues discovered that phytoplankton in regions like the Southern Ocean lack iron,[52] an essential nutrient needed by all organisms. Phytoplankton don't need a lot of the metal, yet as Martin found out, iron concentrations in Southern Ocean surface waters are very low, too low for vigorous phytoplankton growth and a strong biological pump. Iron comes from continents, but Antarctica is covered with ice, so the Southern Ocean depends on iron carried by dust blown in with the wind from distant continents. Because of low iron, today major nutrient concentrations are high at the surface and the biological pump is relatively weak in the Southern Ocean.

That changes when more iron makes its way into the Southern Ocean, potentially plunging the entire planet into an ice age. The complete explanation for ice ages starts with the "Milankovitch cycles,"[53] named after the Serbian geophysicist and astronomer Milutin Milanković who in the 1920s hypothesized that the amount of sunlight hitting the Earth goes through three overlapping cycles that last 23,000, 41,000, and 100,000 years. The variation in sunlight during the Milankovitch cycles, however, was too small to alter the planet's heat budget enough to bring on an ice age, yet it was big enough to set in motion other, more forceful changes in the climate system. One change may have been an expansion of deserts in Asia, sources of iron-rich dust which wind could carry to distant seas,[54] including the Southern Ocean. With more iron, phytoplankton could take advantage of the high major nutrients in the surface layer and primary production would rise. As the biological pump strengthened, atmospheric carbon dioxide declined, and Earth entered an ice age. Although there are other parts to the Southern Ocean hypothesis, such as variation in ocean circulation, the biological pump and its microbial participants are likely crucial for understanding the onset of the last ice age, which ended about 10,000 years ago, and potentially others that waxed and waned over a million years. Some of the biggest swings in the planet's climate over geological time can be linked to what happened to microbes in the ocean.

Since the last ice age, however, oceanic primary production and the biological pump have not changed enough to alter atmospheric carbon dioxide levels and Earth's climate. What about the future? If oceanic primary production is indeed decreasing as suggested by some studies, one hypothesis is that the biological pump will weaken in the coming decades, leading to less carbon sequestered in deep waters and more carbon dioxide left in the atmosphere.

The Biological Pump in the Future

One recent modeling study led by Jamie Wilson explored what the biological pump may look like over the rest of this century.[55] The first part of the study considered a major component of the biological pump, the sinking of organic carbon from the surface layer to deep waters, what biological oceanographers call export production. As expected from evidence showing a decrease in primary production, the models predict a decrease in export production by as much as 1.44 Pg C each year. Wilson and his colleagues attribute the decrease to global warming raising the temperature of the surface layer, leading to lower nutrient levels and lower primary production (Fig. 4.4). As the surface layer warms, the change in water density with depth increases, stratification becomes stronger, and the mixing of nutrient-rich deep waters with nutrient-poor surface waters lessens. As a result, primary production and export production decrease. Acting alone, that should result in a weaker biological pump and less carbon sequestered in deep waters.

But there is more to the biological pump than just export production. Sequestration of carbon by the pump also depends on how long carbon dioxide–laden deep water stays out of contact with the atmosphere, its residence time. Wilson's modeling work argues that the residence time increases with warmer surface waters and stronger stratification—the same

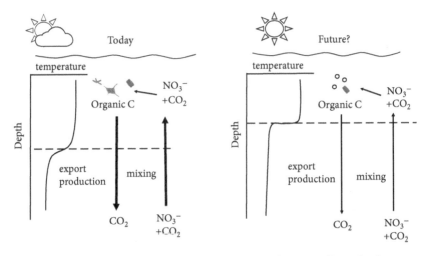

Figure 4.4 Export production and mixing today and potentially in the future. Mixing brings up nutrients such as nitrate (NO_3^-) and carbon dioxide from deep waters to the surface layer. Mixing is inhibited by the warming of surface waters, leading to a steeper gradient in temperature and a shallower mixed layer (above the dashed line). With less mixing, nutrient concentrations would be lower in the surface layer, so primary production and export production would decrease. But less mixing also means less carbon dioxide would be brought up from deep water. In the future, with lower nutrient concentrations, the phytoplankton community may shift to smaller species.

mechanism that explains the decrease in export production. So stratification has opposing effects on the two factors controlling carbon sequestration in deep waters. With stronger stratification, residence time lengthens, and carbon sequestration increases. At the same time, stronger stratification should lead to a decrease in export production and less carbon sequestration. In the battle between residence time and export production, the models indicate the winner is residence time. Wilson and colleagues predict that even though export production will decrease, carbon dioxide sequestration will increase by 19 to 48 percent over the twenty-first century. That amounts to 5 to 17 percent of the total carbon stored over this century and of the carbon dioxide kept out of the atmosphere.

Other effects of a warming ocean may be less beneficial. A reduction in the nutrient supply caused by stronger stratification may select for different types of phytoplankton, with potentially negative consequences for organisms higher up the food chain. Low nutrients favor small phytoplankton like cyanobacteria because they outcompete large phytoplankton like diatoms. As the oceans warm, the regions now dominated by cyanobacteria and other small phytoplankton, such as the subtropical North Pacific, may become even bigger than they already are. There is already evidence that these "ocean deserts" are expanding.[56] Without adequate nutrients, diatoms and other large phytoplankton are rare in ocean deserts, biological production is low, and fisheries are unproductive. Even if nutrients are adequate, higher temperatures could select for undesirable phytoplankton.[57] Warmer waters closer to shore may stimulate the growth of cyanobacteria that produce toxins harmful to other marine biota and even humans on land. Other types of harmful algal blooms may also become more prevalent with global warming.

Phytoplankton may also be challenged by another manifestation of climate change. Along with higher temperatures, the oceans are becoming more acidic, the result of more and more atmospheric carbon dioxide diffusing into surface waters. Chapter 6 will explore how increasing acidity affects phytoplankton, other organisms, and the oceanic carbon cycle.

To sum up, oceanic phytoplankton are responding to rising atmospheric carbon dioxide but in ways unlike what's happening with land plants. While the rise has stimulated more growth by land plants, preventing carbon dioxide from reaching even higher levels, it has not stimulated phytoplankton growth. If anything, climate change appears to have negative effects on phytoplankton. Phytoplankton growth and biomass in the ocean may be decreasing, with even larger drops potentially coming. The "greening of the Earth" is about what's happening on land (Chapter 3), not in the ocean. More carbon from the biological pump may be stored in deep waters, because a decrease in mixing has kept the carbon out of contact with the atmosphere for longer, not because export production has increased with climate change. Along with lower productivity, climate change may select for smaller phytoplankton less suitable for initiating the biological pump and for supporting marine food chains and marine life we most value and appreciate. Disruption of the smallest organisms in the oceans would have consequences as large as those caused by droughts and wildfires brought on by climate change on land.

Phytoplankton and the uptake of carbon dioxide are on one side of the oceanic carbon cycle. Climate change is also altering the other side: respiration and the release of carbon dioxide by oceanic microbes.

Ocean Deoxygenation and the Future of Respiration

As levels of one gas, carbon dioxide, increase in the ocean, levels of another one, oxygen, are decreasing. The ocean has always had oxygen minimum zones (OMZs), where oxygen drops low enough that only bacteria and protozoa can survive for long. The problem is, OMZs are increasing in size and their oxygen levels are dropping even lower.[58] Perhaps even more worrisome, oxygen levels in seawater outside of OMZs are also declining, part of what oceanographers call ocean deoxygenation. Dissolved oxygen has declined in the global ocean by 2 percent over the last 50 years, while in some waters near OMZs the loss is 10 times greater. Even if some fish and invertebrates do fine on lower-than-normal oxygen levels, many organisms are stressed if not killed by any decrease in oxygen. Unlike air, water cannot hold much oxygen so any decrease can have serious consequences. (Coastal marine waters with low oxygen levels, known as "dead zones," differ from those in the open ocean and are discussed in Chapter 8.)

The ocean is losing oxygen because of global warming. As seawater warms, it can hold less oxygen and other gases, the reason why carbonated beverages lose their fizz when warmed up, as mentioned before. The lower solubility of oxygen in warm water accounts for all of the oxygen lost from the upper 100 meters of the ocean,[59] but for the entire ocean, it explains only about 20 percent.[60] Part of the other 80 percent is how global warming is altering winds and currents that mix surface and deep waters. Being in contact with the

atmosphere and sunlight that drives oxygen-producing photosynthesis, surface waters hold more oxygen but fewer nutrients than what deeper waters
hold. The mixing of surface water down to deeper waters, or downwelling,
injects oxygen into oxygen-deficient water, whereas upwelling brings
up deep water with low oxygen but lots of nutrients. The nutrients stimulate algal growth and production of organic matter, some of which eventually sinks via the biological pump to deep water where it is decomposed
by oxygen-consuming bacteria. The net effect of changes in mixing differs
among oceanic regions and is a complicated puzzle being worked out by
oceanographers.

   Another reason why the ocean is losing oxygen is due to the warming
effect on bacteria. Because temperatures are increasing, bacteria use more
oxygen to degrade more organic matter. There are several mathematical
expressions to describe how a chemical reaction, or any process varies with
temperature, some dating back to the work of the Dutch physical chemist
Jacobus Henricus van 't Hoff in the nineteenth century.[61] Ecologists and
microbiologists are more familiar with the Swede Svante Arrhenius and his
eponymous equation, which gives a biological reaction rate as function of
temperature and activation energy, a key property of chemical reactions. But
what ecologists use even more frequently is Q_{10}; it's the change in a rate with
a 10-degree change in temperature. As a rule of thumb, Q_{10} is two. That is,
if the temperature goes up by 10 degrees Celsius, the rate doubles. The first
person to give a version of the Q_{10} rule of thumb was van 't Hoff, although
what he actually said was rates generally "roughly double or triple" with a
10-degree higher temperature.[62] (A tripling of the rate means Q_{10} is equal
to three.) There is no theoretical reason why Q_{10} must be two, yet when it
differs substantially from two, ecologists take notice. A global analysis of
how soil respiration varies with temperature turned up high Q_{10} values, between 2.6 and 3.3, indicating the great sensitivity of soil microbes to temperature.[63] Of the few experiments with marine microbes, most have yielded
Q_{10} values around 2. However, a much higher Q_{10}, equal to 3.6, was derived
from the variation in oxygen consumption rates estimated from changes in
oxygen levels as water masses mixed.[64] Assuming the high Q_{10} value is correct, oxygen consumption by bacteria would increase by nearly 30 percent if
the oceans warm by 2 degrees Celsius. The stimulation of bacterial activity by
warming temperatures is another reason why the oceans are losing oxygen.

   As discussed for soil microbial respiration, temperature-enhanced respiration by bacteria has serious implications for the carbon cycle in the ocean,

marine food chains, and carbon sequestration. Respiration could be higher without a change in DOC uptake if bacteria divert carbon away from synthesis of cellular components toward respiration. In that case, more carbon would be lost from the microbial loop as carbon dioxide and less would be passed onto larger organisms higher up in the food chain. Fish and marine mammals may be affected. Another possibility is that temperature-enhanced respiration is supported by additional use of labile DOC. Use of this DOC can be stimulated by temperature only if production of it is also stimulated; levels of labile DOC are too low to support higher consumption without higher production. The biggest concern is that bacteria would degrade the more refractory parts of the DOC pool, leaving less for export to deep waters or to be stored by the microbial carbon pump. Facing an analogous question on land, many soil microbial ecologists have concluded that global warming is causing degradation of soil organic matter that had previously sequestered carbon away from the atmosphere for hundreds to thousands of years (Chapter 3). Something similar may be happening in the ocean. A study in the North Atlantic Ocean argued that warming stimulates use of refractory DOC components more so than use of labile components.[65]

A warming ocean may not only speed up degradation of DOC but also of zooplankton fecal pellets and marine snow falling through the water column. Degradation of those organic particles leaves less carbon to be exported to deep waters, weakening the biological pump. A modeling study using a Q_{10} of 2 concluded that climate change won't lower the biological pump by much, less than a percent.[66] Assuming a higher Q_{10}, however, translates into a bigger reduction in the pump and less carbon dioxide taken up by the ocean; estimating how much less requires more modeling work. Degradation of sinking organic particles was considered by the study discussed before by Wilson and colleagues who used models to explore the future of the biological pump.[67] Their models included a "transfer efficiency," defined as how much organic carbon in marine snow and fecal pellets survives the journey through the water column from the surface to deep waters. Compared with export production and residence time of carbon in deep waters, the two other components of the biological pump that may change in the future, Wilson was most unsure about how much the transfer efficiency would decrease with global warming. The answer could say a lot about carbon storage in the ocean if climate change continues unabated.

Despite all the possible dire consequences of a warmer, more acidic ocean on marine biota and the carbon cycle, the Intergovernmental Panel on Climate Change (IPCC) is sanguine about the ocean continuing to act as a carbon sink. The latest report from the IPCC says the ocean sink for carbon dioxide has grown over the last six decades, and it seems confident the sink will continue to grow in the near future.[68] Yet there are many reasons to be concerned. The ocean was microbial before climate change. With climate change, it may be become even more so. Ocean deserts populated by small cyanobacteria are likely to expand, and more carbon may be shuttled through the microbial loop and viruses as the oceans warm and become more acidic. Carbon routed through these microbial pathways means less carbon to support large marine life and to funnel into the biological pump. Even if climate change doesn't alter the ocean's contribution to carbon sequestration, it seems likely to disrupt microbial communities that are the base of all life in the ocean.

5

Clouds, CLAW, and a Greek Goddess

Predictions about Earth's climate are based on millions of data points fed into complex models consisting of complicated equations, written in dense computer code that must be run on large, powerful supercomputers. Yet, amid that impenetrable thicket of data, equations, and code, one number stands out: the increase in temperature resulting from a doubling of atmospheric carbon dioxide, what's called the climate sensitivity. Its value determines how long we have before global temperatures rise by more than the target of 1.5 degrees Celsius if we continue to burn fossil fuels at the current rate. The fear is if the climate is more sensitive than we now think, then the window for preventing ruinous global warming is much smaller. There is still much uncertainty about the climate sensitivity value, despite having been examined since the late nineteenth century.

One estimate was published in 1896 by Svante Arrhenius, the same Swede known to microbial ecologists for his eponymous equation that gives the rate of a reaction as a function of temperature (Chapter 4). At the time, Arrhenius was working on other calculations that have made him well-known to climate-change scientists, five years before he was finally elected to the Royal Swedish Academy of Science after much opposition and seven years before he would receive the Nobel Prize in Chemistry for his work on acid-base chemistry. As the marriage to his first wife and former laboratory assistant, Sofia Rudbeck, was coming to an end, Arrhenius busied himself with calculating how much Earth would warm as atmospheric carbon dioxide increases.[1] His initial calculations indicated that temperatures would rise by 5 to 6 degrees Celsius if carbon dioxide levels doubled,[2] and then later he revised his estimate down to 4 degrees Celsius. His lower number is close to the current best estimate of 3 degrees Celsius, calculated by much more sophisticated means.[3] In addition to being the first to estimate Earth's climate sensitivity, he was among the first to recognize that carbon dioxide and water vapor are greenhouse gases.

What dims the luster of Arrhenius's status as a global warming prophet is that he thought more carbon dioxide in the atmosphere and the warmth it would bring would be good. He did the calculations to test his hypothesis that a large reduction in carbon dioxide concentrations would lead to lower global temperatures and bring on glaciation. He and several other scientists at the time feared the coming of another ice age that, as Arrhenius put it, would force people to move from "our temperate countries into the hotter climates of Africa."[4] But he wasn't too concerned because "the enormous combustion of coal by our industrial establishments suffices to increase the percentage of carbon dioxide in the air to a perceptible degree." Arrhenius thought more carbon dioxide would benefit a "rapidly propagating mankind" and suggested:

> By the influence of the increasing percentage of carbonic acid [carbon dioxide] in the atmosphere, we may hope to enjoy ages with more equable and better climates, especially as regards the colder regions of the earth, ages when the earth will bring forth much more abundant crops than at present, for the benefit of rapidly propagating mankind.[5]

Arrhenius's home in Uppsala, Sweden, was in one of those cold regions where some warming may raise crop yields and lower heating costs. Arrhenius did not foresee that we would burn so much fossil fuel that our climate would become less equable and far from being better.

In the over 125 years since Arrhenius did his calculations, climate scientists have learned a lot more about climate sensitivity. As described in the latest report from the Intergovernmental Panel on Climate Change (IPCC), the range of estimates has narrowed over the last 30 years, from 1.5 to 4.5 degrees Celsius in 1979 to 2.5 to 4 degrees Celsius today (the "likely" estimates).[6] The tightening of the range is a sign of progress, yet the remaining gap is also a sign of work remaining to be done. Some modeling work suggests the climate sensitivity could be as high as 5.6 degrees.[7]

One reason for the uncertainty about climate sensitivity is aerosols. Climate models yield different estimates because they differ in how they deal with small, airborne particles and the clouds they may make. Aerosols directly contribute to Earth's heat balance by reflecting sunlight back to space, and they can indirectly contribute by helping to make clouds. Clouds form when warm, moist air rises.[8] With increasing altitude in the atmosphere, air cools and the pressure drops, forcing water vapor to condense into small droplets and ice crystals, eventually forming clouds. But unless temperatures drop far below freezing or the air becomes supersaturated with water vapor, clouds will not form in the absence of cloud seeds, or cloud condensation nuclei (CCN), which are small particles that water vapor condenses on to yield clouds. Most cloud seeds are between 0.04 and 0.3 microns in diameter, about the size of viruses, way smaller than the 70-micron diameter of a human hair. Larger particles can form clouds, but they fall quickly with rain, making them too rare to contribute much.

Over land the particles can be dust, clay, and soot from forest fires, smokestacks, and car tail pipes, or organic chemicals from human-related and natural sources. Over most of the planet—that is, over the ocean—many particles come from seawater. And many of those particles can be traced back to marine microbes.

The microbe–cloud connection seems more tangible than the other ways that microbes shape our climate. Most of the ways discussed in this book aren't visible to the naked eye or even with the most powerful electron microscope. We must rely on scientific arguments presented in dense publications showing how microbes change greenhouse gas levels and how these gases set the planet's heat budget. Clouds are more visibly linked to climate. We can readily appreciate how clouds shade out the sun on a hot summer day and cool us down. Yet the story of how microbes contribute to the formation of clouds is as complicated as the links between microbes and greenhouse gases. The story is populated by colorful scientists who compiled compelling evidence for an imaginative hypothesis named after a Greek goddess. Even though the most audacious part of the hypothesis was shot down, the story is still worthwhile to recount if only to learn more microbial ecology and biogeochemistry and to illustrate how science is done. What remains accepted today is another example of how microbes, the unseen agents, shape Earth's heat budget and climate.

Clouds from Microbes

James Lovelock had problems finding the ship that was supposed to take him to Antarctica in November 1971.[9] Because the RRS *Shackleton* was so small, her mast and infrastructure were below the docks at Barry, Wales, and could be seen only when Lovelock was right at *Shackleton*'s berth. The ship was tiny for such a long voyage from the United Kingdom to the Southern Hemisphere, but Lovelock thought there was less contamination on small vessels than on larger ones, which he called "floating palaces." (I wonder where he got that notion. Oceanographic ships of all sizes are hardly luxurious.) Using "housekeeping money" "willingly donated" by his first wife, Helen, after British funding agencies turned him down, Lovelock's goal was to measure chlorofluorocarbons (CFCs) in pristine air using the gas chromatography method he had invented[10] to test hypotheses about the transport and fate of CFCs. Lovelock's data collected on the *Shackleton* were instrumental in work by M. J. Molina and F. S. Rowland, who demonstrated that CFCs destroy ozone and reduce protection from DNA-damaging ultraviolet sunlight.

Lovelock also wanted to measure another gas, dimethyl sulfide (DMS). He was familiar with DMS from his early experiments in the mid-1950s to determine what attracted flies to rotting seaweed piled onto the beaches of southern England. As one of several chemicals that give a seashore its characteristic smell, DMS is used by marine birds like albatrosses, petrels, and shearwaters to find productive waters and possible food.[11] Years after his rotting seaweed work, Lovelock was looking for a mechanism by which sulfur moves from the ocean to the land, thereby closing the global sulfur cycle. Without this mechanism, land would become deficient in sulfur as it drains into the ocean, stunting the growth of terrestrial organisms. All organisms need a little sulfur for a couple of amino acids and a few other essential biochemicals. Sulfur would be restored on land if the oceans released another sulfur gas, hydrogen sulfide, but it is produced by bacteria only when oxygen is absent, not in surface waters of the oceans where oxygen is abundant. DMS seemed a more likely candidate to close the sulfur cycle. Lovelock's first paper on the topic reported that DMS levels in the Atlantic Ocean were much higher than in soils from several sites in the United States and that more of the gas came from marine algae than from leaves of cottonwood, spruce, oak, and pine trees.[12] At the end of the paper, he points out that degradation of DMS in the atmosphere produces sulfuric acid and other sulfur chemicals.

Lovelock didn't realize the significance of those DMS degradation end-products until over 10 years later when in 1986 he was invited to spend a month at the University of Washington in Seattle. Discussions with his host, Bob Charlson, turned to the topic of cloud formation. Lovelock was surprised to learn clouds need cloud seeds, those CCN, to form, and he was even more surprised to hear that the source of CCN over most of the planet—the ocean—wasn't known. Many particles that could seed cloud formation were known to contain sulfuric acid (H_2SO_4) or sulfate (SO_4^{2-}), but where those sulfur chemicals came from was a mystery. The known sources such as volcanos and polluting industries are too far away from most of the world's oceans. The source had to be local. One possibility was Lovelock's sulfur gas, DMS, produced by marine algae, including common phytoplankton in the ocean.

Charlson and Lovelock teamed up with two other scientists, M. O. Andreae and S. G. Warren, to present their ideas about DMS and cloud formation in one of the most influential papers in environmental science.[13] Published in 1987 yet still quoted today, the paper put forward what is often called the CLAW hypothesis, an acronym based on the authors' last names. The first part of CLAW argued that sulfuric acid and sulfate aerosols are the main source for CCN over the open ocean far from pollution and volcanos, and that those sulfur chemicals come from DMS. Once in the atmosphere, DMS reacts with the hydroxyl radical to form methane sulfonic acid or sulfur dioxide, which can be further transformed to sulfuric acid. All three sulfur chemicals can contribute to particles of the right size to act as CCN. The CLAW authors went on to discuss how changing the size and number of CCN affects cloud albedo, which is the capacity of clouds to reflect sunlight back into space. They hypothesized that production of DMS by marine phytoplankton would increase cloud cover, reflect more sunlight, and act to cool the planet, thereby helping to negate global warming caused by rising atmospheric carbon dioxide. Recent calculations indicate that a 20 to 35 percent increase in low clouds could neutralize a doubling in carbon dioxide levels.[14] In effect, the CLAW hypothesis said microbes could slow down global warming.

We now know a lot more about DMS and what CLAW got right—and wrong. As mentioned in the 1987 CLAW paper, the production of DMS is complicated and involves many microbes, not just phytoplankton. The precursor of DMS is another sulfur chemical, dimethylsulfoniopropionate (DMSP), which is made by several marine phytoplankton. Algal physiologists have suggested several reasons why these phytoplankton make DMSP, including for use in the synthesis of other cellular components, to serve as an antioxidant or for cryoprotection, or to fend off predators. According to the most widely accepted hypothesis, however, DMSP is an osmolyte that phytoplankton use to cope with high salt concentrations. Regardless of DMSP's function, phytoplankton can produce DMS from it, but more commonly they release DMSP into seawater or DMSP spills out during predation by zooplankton (Fig. 5.1). The now dissolved DMSP is consumed by bacteria.[15] Usually, bacteria demethylate DMSP to form 3-methiolpropionate, but they

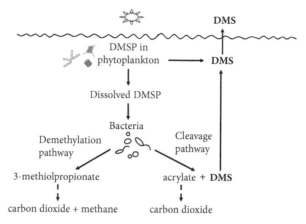

Figure 5.1 The production of dimethyl sulfide (DMS) from the degradation of dimethylsulfoniopropionate (DMSP). Based on Moran and colleagues (2012).

can also cleave it to produce acrylate and DMS. So production of DMS is complicated. Not only are many organisms and chemical reactions needed to make DMS, but also DMS is just one of several byproducts of DMSP degradation. These complexities need to be kept in mind when we look at the connection between phytoplankton and clouds.

The best evidence for the first part of the CLAW hypothesis—that DMS degradation releases cloud seeds (CCN)—comes from Down Under. Perched on the northwest tip of Tasmania south of Australia, the research station at Cape Grim samples gases and other air-borne constituents carried by prevailing westerly winds sweeping over the Southern Ocean. In 1976, the station started to measure atmospheric carbon dioxide from an old NASA caravan and continues today to collect invaluable data about greenhouse gases in the Southern Hemisphere.[16] Because the station is so remote, far from volcanos and anthropogenic sources of cloud-forming particles, atmospheric scientists also have been able to explore connections between CCN and DMS.[17] They found that CCN, DMS, and two byproducts of DMS degradation, methane sulfonic acid and sulfate, were high in summer and low in the winter; they follow the activity of DMS-producing phytoplankton like *Phaeocystis* in the Southern Ocean where some of the highest concentrations of DMSP have been measured. Similar studies in the tropical South Atlantic and the northeast Pacific Ocean also found a high correlation between CCN and DMS. The correlation is a good argument for the first part of the CLAW hypothesis.

Microbes as Goddesses

The other, more controversial part of the CLAW hypothesis evoked a Greek deity. After proposing that cloud seeds come from DMS made by marine microbes, the CLAW authors went on to hypothesize that these microbes, especially DMSP-rich phytoplankton, respond to changes in temperature and sunlight by releasing more DMS, which in turn leads to more clouds and cooler temperatures. If cloud cover becomes too thick, DMS production slows down, according to CLAW, and cloud formation diminishes. In effect, marine microbes act as a "planetary thermostat"[18] that controls Earth's temperature and thus climate. The authors linked this control mechanism to the Gaia hypothesis, named after a primordial goddess and the mother of all life in Greek mythology, as mentioned in Chapter 1. James Lovelock, the "L" in CLAW, had first proposed Gaia in 1972 in a short letter to the editors of an academic journal.[19] He got the name "Gaia" from his friend and neighbor, William Golding, who won the Nobel Prize in Literature in 1983 and is best known for his debut novel, *Lord of the Flies*. Lovelock eventually recognized "Gaia" was better than alternatives such as "Biocybernetic Universal System Tendency/Homeostasis."[20]

Gaia was conceived in California in the mid-1960s. Lovelock was invited by NASA in 1961 to work at the Jet Propulsion Laboratory (JPL) in Pasadena, California, as part of the space agency's mission to explore the Moon.[21] The offer was a dream come true for Lovelock. As he said to his biographers, "I was being asked to join in the kind of adventure that just a few years back I had been reading as science fiction."[22] But he soon became bored by the Moon project. His enthusiasm returned when he started to work on another project, the search for life on Mars. While working on that project in September 1965, Lovelock had the epiphany that spawned Gaia. He learned that the composition of the Red Planet's atmosphere was dominated by carbon dioxide, very unlike Earth's atmosphere, which has only very low levels of carbon dioxide but lots of nitrogen and oxygen and substantial amounts of methane (Fig. 5.2, data from Lenton[23]). For Lovelock, the difference between the two planets answered the question about life on Mars. After looking at the composition of its atmosphere, Lovelock knew the planet was lifeless. Unlike Earth, Mars must not have the biology—the microbiology—that produces or consumes the four gases. Lovelock then realized the implications of biology setting a huge planetary feature, the composition of Earth's atmosphere, which in turn determines the planet's climate. He discussed his idea with an astronomer

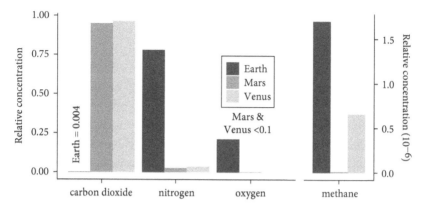

Figure 5.2 Concentration of four gases on Earth, Mars, and Venus. The differences between the three planets demonstrate the impact of biology— mainly microbiology—on the atmosphere and thus climate. Data from Lenton (1998).

working at JPL at the time, Carl Sagan, who thought Gaia would explain how Earth's temperature has been maintained (more or less) over billions of years even though the Sun's output has increased; after being at first low when Earth came into existence, it has intensified by 30 percent ever since, over the last four billion years. (Sagan testified before the U.S. Congress in 1985 about global warming,[24] and he became well-known at the time for his many popular science books, his 13-part television series *Cosmos*, and his science-fiction novel *Contact*, which in 1997 was turned into a movie of the same name.) Sagan suggested that Lovelock talk to his former wife, Lynn Margulis.

About the time Lovelock and Margulis first met, she was an unknown microbiologist at Boston University struggling to get her ideas about microbial evolution accepted. She was trying to explain one of the biggest quantum leaps in evolution, second only to the origin of life itself: how eukaryotes arose from prokaryotes. The differences between plants and animals pale in comparison to the chasm separating prokaryotes, that is, bacteria and archaea, from all other organisms, the eukaryotes. As mentioned in Chapter 1, the prokaryotes don't have a nucleus to house their genome, nor do they have any other organelles like chloroplasts, the site of photosynthesis, or mitochondria, where power to run cell metabolism is generated. All eukaryotes have a nucleus and most have mitochondria, whereas plants and algae have chloroplasts. Building on long dormant ideas proposed by Russian botanists and others early in the twentieth century, Margulis argued that organelles in eukaryotes were originally bacteria that had been ingested but not digested by a proto-eukaryotic host. The bacteria became symbionts inside of the host, or endosymbionts, providing novel metabolisms to help both symbiont and host survive and proliferate. The idea came to be called the endosymbiosis hypothesis. Margulis's first paper on the topic in 1967 was supposedly rejected by 15 journals before it was finally published,[25] yet today the hypothesis is widely accepted. It is now thought that the chloroplast evolved from a cyanobacterium while the mitochondrion was a type of heterotrophic bacterium in the Alphaproteobacteria phylum. One hypothesis says the original host cell was an archaeon.[26] Although much remains unknown, the endosymbiosis hypothesis is the best explanation biologists have for the origin of eukaryotes.

Margulis's expertise in microbiology breathed life into Lovelock's Gaia hypothesis. Even though (or perhaps because) Lovelock's PhD was in medical science,[27] he had considered bacteria to be only pathogens, not powerful agents in the biosphere that potentially could modify Earth's climate by producing and consuming greenhouse gases. Two years after Lovelock published his short note announcing the Gaia hypothesis, he and Margulis put forth in 1974 two much more substantial papers that fleshed out how Gaia would work. The order of author names switched from Lovelock and Margulis on one paper[28] to Margulis and Lovelock on another.[29] Probably because of Margulis, the papers discuss how microbes produce and consume several gases in Earth's atmosphere, including nitrogen, methane, nitrous oxide, and ammonia. Lovelock and Margulis went on to argue that during the first billion years or so of Earth's existence, life had to modify greenhouse gas levels in the atmosphere to maintain the planet's temperature within a livable range. As they put it:

> A first task of life was to secure the environment against adverse physical and chemical change. Such security could only come from the active process of homeostasis in which unfavorable tendencies could be sensed and counter measures operated before irreversible damage had been done.[30]

"Homeostasis," mentioned in the quote here and used in the title of the Lovelock and Margulis 1974 publication, is the heart of Gaia. From its beginning, according to the Gaia hypothesis, Earth's biota has been a self-regulating system that has kept the planet habitable, even optimal for life.

Lovelock and Margulis's papers were largely ignored for the first 10 years after their publication, as were the three other academic papers on the topic they wrote together in the 1970s. In his book-long critique of Gaia,[31] Toby Tyrrell argued that the idea didn't attract much attention from scientists until 1979 when Lovelock brought out his book, *Gaia: A New Look at Life on Earth*. The book was influential despite the lack of citations to scientific publications and its breezy, informal style; Lovelock said his goal was "to stimulate and entertain."[32] Even after Lovelock's book came out, however, few publications mentioned Gaia or even cited the Lovelock-Margulis papers. Unlike nearly all other scientific publications, the Lovelock and Margulis paper has been cited much more frequently in the last 10 years than it was during the first 10 years of its existence, 281 and 28 citations, respectively.[33] Of the papers citing Lovelock and Margulis before 1986, most didn't even mention "Gaia," much less take on the hypothesis in any detail. Readers of those papers could be forgiven if they missed the brief nod to Lovelock and Margulis. By contrast, readers of the CLAW paper, published in 1987, couldn't avoid Gaia. The paper's evocation of the goddess is a big reason why it has been so consequential in climate science and beyond.

The CLAW authors opined that DMS release by marine microbes was Gaia at work. After reviewing data and calculations about DMS contributing to cloud seeding, the authors went on to argue that DMS production and its link to cloud formation could be a negative feedback mechanism that keeps temperatures tolerable for life (Fig. 5.3). Unlike positive feedback (Chapter 3), negative feedback keeps climate change from getting worse. If temperatures or sunlight or some combination of both increases, DMS production is stimulated and clouds become more numerous, forcing temperatures to decline. Temperatures would not get too cold because DMS production should decline as marine waters cool and clouds block the sunlight needed by DMSP-producing phytoplankton. Since publication of the CLAW hypothesis over 35 years ago, several studies have provided evidence to argue for or against the feedback mechanism.

One of the strongest pieces of evidence for the feedback mechanism came from a study that looked at light intensity and DMS concentrations in the surface ocean.[34] As part of his PhD dissertation working with Rafel Simó at the Institut de Ciències del Mar in Barcelona, Sergio Vallina looked at the relationship between light and DMS in surface waters in the Mediterranean Sea near Blanes, Spain, and at Hydrostation S in the Sargasso Sea off Bermuda. Vallina and Simó also took data where they could from around the world. In

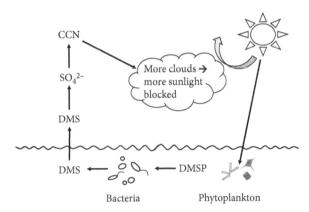

Figure 5.3 The CLAW hypothesis. A rise in temperature and light stimulates dimethylsulfoniopropionate (DMSP) production and release of dimethyl sulfide (DMS). Once in the atmosphere, DMS is oxidized to sulfate (SO_4^{2-}), which forms cloud condensation nuclei (CCN) and ultimately clouds that block sunlight and cool the planet.

all data sets, they found that light intensity explained a large fraction, over 90 percent, of the variation in DMS levels over time. As predicted by CLAW, more light leads to more DMS. Vallina and Simó used two arguments against an alternative hypothesis that the apparent stimulation of DMS production by light is just a byproduct of vigorous phytoplankton activity; with more light phytoplankton are more active and perhaps produce more DMS. First, in these surface waters, phytoplankton growth is set by nutrients, not light, so a higher intensity wouldn't necessarily increase phytoplankton activity in surface waters. Second, the relationship between DMS and a measure of phytoplankton biomass, chlorophyll, was negative, not positive as would be expected if general phytoplankton activity determined DMS production. Light may stimulate DMS production via how it affects production of its precursor, DMSP. Even though light is not required to synthesize DMSP, a phytoplankton cell may make more of DMSP for protection against the damaging effects of ultraviolet radiation. Other studies have also found a positive relationship between DMSP levels and light intensity.[35]

So the connection between DMS and light intensity could contribute to the negative feedback mechanism needed by Gaia for DMS and microbes to be part of a planetary thermostat. As sunlight increases, DMS from marine microbes may counteract the heat brought in with it. More light leads to more DMS and CCN that form more clouds. A theoretical model suggested that without DMS from marine microbes, fewer clouds form and temperatures warm up.[36] If so, DMS emissions by marine microbes exemplify Gaia; by making clouds, the biota maintains homeostasis, keeping temperatures comfortable and Earth habitable.

The Case against CLAW

The first attacks hit the foundation of the hypothesis, that DMS is the main source of seeds necessary to form clouds over the oceans. Several studies have found additional sources of CCN. Sea salt thrown up into the air by breaking waves could be one.[37] Sea salt can account for up to 65 percent of CCN under some atmospheric conditions in the high latitudes of the Southern Hemisphere, but generally it contributes less than 30 percent over the global ocean. More important may be organic chemicals other than DMS. According to work in the Atlantic and Pacific Oceans, the organic chemicals are common, often the biggest part of marine aerosols, which reflect sunlight and can contribute to cloud formation.[38] The organics in aerosols include the chemicals produced during the degradation of organic gases released from the ocean. Along with DMS, the gases include the terpenes isoprene and monoterpene.[39] Like sea salts, larger dissolved organic chemicals could contribute to CCN when they are injected into the air by breaking waves. Some possibilities for the organics are polysaccharides and proteins recently made by phytoplankton and other microbes as well as the very old, refractory dissolved organic carbon described in Chapter 4. Sophisticated chemical analyses such as mass spectrometry, Fourier transform infrared spectroscopy, and Raman microscopy have revealed the organic matter coating the surface of salt particles. Microbes can also become airborne by wave action. Small algae, most bacteria, and nearly all viruses are the right size to act as CCN, or their cellular components could do the job. These other sources of CCN are one reason why modeling studies have found the connections between DMS, CCN, and carbon dioxide levels to be weak.[40] One study found that a 50 percent rise in carbon dioxide would stimulate only a 1 percent rise in DMS and an increase in CCN of only 0.1 percent. Sources of CCN other than DMS aren't part of CLAW.

An important part of CLAW is the connection between DMS and temperature. If DMS is to be Earth's thermostat, it needs to respond to temperature to earn its name. Although there is a strong correlation between DMS and sunlight, the connection to temperature is less clear. The CLAW authors pointed out that emission of DMS into the atmosphere was greatest in the tropical and equatorial oceans, which seemed true according to data available at the time. The data supported the authors' statement that "the most important climatic role of DMS is to contribute to elevated cloud albedo over the warmest ocean regions, and thus to reduce the input of heat into the low-latitude oceans."[41]

However, DMS measurements taken since CLAW was first published argue against this part of the CLAW story. Rather than being highest in the warmest waters, DMS levels are highest in cold waters of the Arctic Ocean and the Southern Ocean, and unlike the high correlation between DMS levels and light, DMS and temperature are only weakly correlated ($r = 0.18$).[42] Data on DMS emissions, which are calculated from the wind speed and the difference between DMS levels in seawater versus in the atmosphere, also don't help CLAW much. Although there are some DMS hotspots in the Pacific Warm Pool as CLAW would predict, DMS emission is also high in cold waters of the subarctic Pacific, the Arctic, and the Southern Ocean. If Gaia were working and if DMS production were a thermostat, DMS emissions should be low, not high, in these frigid regions. Gaia seems absent.

The variation in DMS levels and emissions among locations is partially explained by the types of phytoplankton growing in different oceanic regions. Most warm marine waters have low nutrient concentrations and are populated by cyanobacteria and small eukaryotic phytoplankton that either don't make DMSP at all (most cyanobacteria) or only small amounts (many small eukaryotic algae).[43] Without its precursor, DMS is low in these waters. Cold oceanic regions, on the other hand, can have high nutrient levels and large populations of two types of DMSP-rich phytoplankton, *Phaeocystis* and coccolithophores. DMS emissions differ even among cold waters, depending on which phytoplankton is present. In one study the presence of *Phaeocystis* explained why air above the Greenland Sea had higher DMS levels than did another Arctic region, the Barents Sea, even though the Barents Sea had higher total phytoplankton biomass.[44] At the other end of the world, *Phaeocystis* is often abundant in the Southern Ocean where DMS emissions can be high. The relative abundance of *Phaeocystis* versus diatoms, a type of phytoplankton that doesn't produce much DMSP and no DMS, seems to be set by temperature, light availability, and iron levels, modulated by predation.[45] To fend off attack by zooplankton, *Phaeocystis* increases in size by 100-fold and changes its morphology, transitioning from 10-micron single cells to mucus-rich colonies that are about a millimeter in diameter. Abundant DMSP-rich *Phaeocystis* helps to explain why DMS levels are high near Antarctica.

High DMS in cold water wasn't predicted by CLAW. The lack of a strong connection between DMS emissions and water temperature and other arguments have led to the demise of the CLAW hypothesis. As one highly cited review put it, "If the CLAW hypothesis has not stood the test of time, it is only because we now have a much better appreciation of the complexity of biogeochemistry and climate physics than when the hypothesis was first put forward."[46] Yet, even if CLAW has been put to rest, important parts of it remain alive. DMS continues to be a major source of cloud seeds over the ocean, and there is no doubt about the importance of CCN and clouds in establishing the planet's heat budget. One problem in more firmly establishing the role of DMS in cloud formation is the paucity of data and the complexity of links between DMS, aerosols, cloud formation, and cloud albedo. As one indication of the problem, marine geochemists don't even agree whether DMS emissions have gone up or down since 1980, although all agree more DMS will be released from polar waters as sea ice melts.[47] Models also indicate that DMS emissions will decrease in the future as global warming leads to stronger stratification and less growth of DMSP-rich phytoplankton.

Another important part of the CLAW hypothesis—the connection between phytoplankton and clouds—is still apparently true; all organic carbon contributing to cloud seeding over the ocean comes from phytoplankton. The number and diversity of CCN sources complicate attempts to link microbes and clouds, and the different contributions by different types of microbes add another layer of complexity. Despite all those complications, however, climate scientists have found several relationships between phytoplankton and clouds, such as the correlation between phytoplankton biomass (chlorophyll) and the light-reflecting tendency of clouds (their albedo) over the Atlantic Ocean[48] and the Southern Ocean.[49] Another study found a high correlation between the abundance of small phytoplankton and the number of CCN over the Mediterranean Sea.[50] Whether via DMS or another organic chemical, microbes seem important in making clouds over vast regions of the planet.

Postscript about Gaia

Just as CLAW has been retired, so too has its grander parent, Gaia, been declared dead. In the last chapter of his takedown of the hypothesis, Tyrrell says, "Gaia is a fascinating but a flawed hypothesis. It is not a correct characterization of planetary maintenance and life's role therein."[51] He accepts two weak forms of Gaia, "influential Gaia" (biota affects the physical and chemical world) and "coevolutionary Gaia" (biota and the physical-chemical world influence each other) but argues there is nothing new about the two ideas, and they aren't really versions of the Gaia theory. He doesn't believe in a stronger form of Gaia, "homeostatic Gaia" in which the biota manipulates the physical-chemical world to maintain conditions stable for life. Just the opposite has occurred over geological time. The evolution of an oxygen-evolving form of photosynthesis around 2.3 billion years ago eventually led to the rise in oxygen in the atmosphere and the near-extinction of many oxygen-intolerant species. Because trees and phytoplankton have been so efficient in drawing carbon dioxide out of the atmosphere, Earth has been plunged into ice ages, the most severe ones known as Snowball Earths, when thick sheets of ice covered much of the planet between 700 and 580 million years ago, snuffing out all life except for a few hardy microbes. Snowball Earths and other, less dramatic ice ages should not have happened if homeostatic Gaia were on watch.

Yet a gentrified form of Gaia remains in what's now called Earth system science: the bringing together of biology, chemistry, and physics (and more) to understand the interconnections between the atmosphere, the biosphere, the cryosphere, the geosphere, and all the other spheres making up the planet. In Earth system science, biology is mostly microbiology. Although CLAW may be tarnished and Gaia has metamorphosed into an academic field, a key concept—the importance of microbes in shaping the planet's atmosphere, its heat budget, and thus climate—remains.

6

Slow Carbon and Deep Time

The White Cliffs of Dover are visible from France on a clear day, thanks to their size, towering to 100 meters in places and stretching nearly 13 kilometers on both sides of the town of Dover, and to the brilliance of the cliff's main rock, chalk.[1] The soft white rock was deposited 100 to 66 million years ago when England and much of Europe were covered by a warm shallow sea. Then starting about 66 million years ago, tectonic forces lifted the chalk deposits above sea level, leaving behind only a ridge connecting England and France. The ridge held back a freshwater lake where today's North Sea is located, until between 450,000 and 180,000 years ago when it was breached by lake water, destroying England's connection to France. As the flood of freshwater cut through the deposits, the chalk was exposed, and the White Cliffs of Dover emerged.

Made of calcium carbonate, chalk is part of the largest stock of carbon on Earth, containing over 100,000 times more carbon than what's in the atmosphere (Chapter 2). Compared to the other carbon stores on the planet, carbonate-rich rocks have a different place in the carbon cycle and in thinking about climate change. Also different are their relationships with microbes. The microbial connection to chalk was discovered by Thomas Henry Huxley, better known as "Darwin's bulldog" for his fervent defense of evolution and the man who proposed the theory. On August 26, 1868, he discussed another topic in a lecture titled "On a Piece of Chalk," given at a meeting of the British Association for the Advancement of Science, held in Norwich, a town about 60 miles north of Dover.[2]

Huxley began his lecture by saying that if workers would dig under Norwich, they would find a layer of chalk extending several hundreds of feet below the surface. The chalk continues to the north as far as Yorkshire, while in the south, "it supplies that long line of white cliffs to which England owes her name of Albion."[3] He went on to point out that England's chalk is connected to vast deposits of the white rock throughout Europe, Asia, and North Africa. Chalk runs west too. Huxley's friend and shipmate, Captain

Dayman, gave him samples from the bottom of the North Atlantic Ocean where a telegraph cable connecting England and America ran. The samples were full of chalk. For good reason the geological epoch when the great chalk deposits formed is called the Cretaceous, after the Latin word for "chalk." Where does all this chalk come from? Huxley claimed, "A great chapter of the history of the world is written in the chalk."[4]

Huxley found the answer when he cut into chalk and trained a microscope onto a thin slice. He saw what he called coccoliths, the term still used today to describe the plates of calcium carbonate made by a type of algae, coccolithophores (Fig. 6.1). Why coccolithophores festoon their outer cell surface with coccoliths is not entirely clear, but one hypothesis is that the carbonate plates protect these algae from attack by zooplankton.[5] As coccolithophores died during the Cretaceous, their carbonate cell walls sank to the bottom, escaped degradation and dissolution, and were compressed to form chalk. The rain of carbonate from coccolithophores to the bottom was light, adding only half a millimeter to the ocean floor each year, equivalent to a few hundred coccoliths, yet enough over millions of years to build deposits up to 500 meters thick in places. Even with only this light rain, many oceanic sediments are rich in carbonate. About half of the carbonate made by coccolithophores and other microbes in the surface layer makes it to the bottom, much more than the 1 percent of organic carbon that survives the journey.[6]

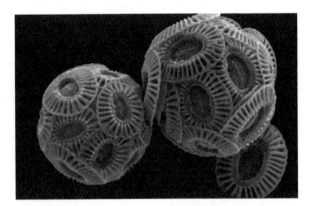

Figure 6.1 The coccolithophore *Emiliania huxleyi* covered with coccoliths. Each cell is about 4 microns in diameter. Picture from Alison R. Taylor, University of North Carolina, Wilmington, Richard M. Dillaman Bioimaging Facility.

In recognition of Huxley's contribution, his name was given to the most abundant coccolithophore species, *Emiliania huxleyi*. The genus name honors Cesare Emiliani,[7] an Italian-American paleontologist who also studied another group of small marine organisms, foraminifera, which also come armored with calcium carbonate shells, called tests by foram experts. Microfossils of foraminifera preserved in oceanic sediments have yielded many insights into the carbon cycle of the geological past and ironically have helped oil companies find fossil fuels.

Coccolithophores are much more numerous and more important in making calcium carbonate, or calcification, than are foraminifera in today's ocean.[8] These algae account for at least half of calcification on the planet, with the rest split between foraminifera, a few other microbes, and pteropods, which are pelagic sea snails that live in the water column. Reef-building corals are larger, more conspicuous calcifiers but cover too little area to contribute much today. By contrast, coccolithophores can form huge blooms blanketing hundreds of thousands of square kilometers for as long as seven months, starting in early spring in both the Northern and Southern Hemisphere.[9] They can account for as much as 40 percent of primary production in the ocean, although their contribution on average is less than 10 percent. Because of their coccoliths, blooms of coccolithophores turn the ocean a milky turquoise, making them visible to satellites overhead, and rapidly growing coccolithophores produce so many coccoliths that some detach from the algal cell and add more color to surface waters. The light color increases the albedo of the ocean and the reflection of sunlight and heat back to space. Coccolithophores, along with diatoms, dinoflagellates, and cyanobacteria, dominate phytoplankton communities of today's ocean.

Coccolithophores vie with diatoms for being the most important phytoplankton in the biological pump, the main mechanism by which atmospheric carbon dioxide is transported to the deep ocean (Chapter 4). If an algal bloom occurs in oceanic waters, it is often dominated by coccolithophores or by diatoms. Both algae take up a large amount of carbon dioxide to make organic material, and their cells are relatively big, leading to the production of big organic particles that sink to deep waters. Also, the hard minerals in the cell walls of coccolithophores and diatoms are ballast that helps organic particles sink faster, ensuring more carbon is transported to deeper waters which remain out of contact with the atmosphere for longer. The cells walls of these algae differ, however. Unlike the calcium carbonate plates of coccolithophores, diatoms use silica for their cell walls. Silica is found in only a few other pelagic organisms, such as silicoflagellates and radiolarians. Although the cell walls of diatoms and coccolithophores may seem exotic, their main constituents appear in familiar materials of everyday life. Calcium carbonate is a main ingredient of cement and silica is in glass. Vast areas of the ocean floor are covered by calcium carbonate or opal (a form of silica) and have been used by paleoceanographers to deduce rates of primary production in the geological past. Just as carbonate from coccolithophores has made chalk, silica from diatoms is the main mineral of another sedimentary rock, chert.[10] Not all carbonate rocks are from coccolithophores, nor is all chert from diatoms. Yet the cell walls of both algae have provided the minerals that make up many rocks now dominating Earth's geology.

Coccolithophores and diatoms both contribute carbon to the organic or soft part of the biological pump, but only coccolithophores and a few other pelagic organisms supply calcium carbonate for the hard part of the pump, also called the carbonate pump. The amount of carbon it transports to deep water is small, less than 10 percent of the complete biological pump,[11] yet it's an important part. The huge amount of carbon deposited in ocean sediments by the carbonate pump is carbon not in the atmosphere. But operating the carbonate pump comes with a cost. The formation of a molecule of calcium carbonate releases one molecule of carbon dioxide as described by the following equation:

$$Ca^{2+} + 2HCO_3^- \rightarrow CaCO_3 + CO_2 + H_2O$$

where Ca^{2+}, HCO_3^-, and H_2O are calcium ion, bicarbonate, and water, respectively. By consuming bicarbonate, coccolithophores also reduce seawater's capacity to buffer against an increase in acidity. So, in terms of climate change, calcium carbonate formation is both detrimental and beneficial, depending on the time frame. It has a negative effect in the short term (it produces carbon dioxide and reduces seawater's buffering capacity), but the storage of carbon in carbonate rocks is a positive over the long haul (it keeps carbon dioxide out of the atmosphere).

Coccolithophores are one connection between the fast carbon cycle and a slow one. The fast carbon cycle, the cycle featured so far in this book, is populated by short-lived organisms, most notably microbes with lifespans measured in hours or days, running reactions that operate over equally short time spans. Depending on the habitat and instrumentation, the production and consumption of carbon dioxide by microbes can be measured in minutes. Living coccolithophore cells are firmly in the fast carbon cycle, yet their coccoliths buried in sediments are part of the slow carbon cycle, at the other end of the time scale. Rather than minutes or even decades, the slow carbon cycle operates over thousands to millions of years, what science writer John McPhee called "deep time."[12] We've already discussed long periods of time and geological ages thousands and millions of years ago. This chapter will have more about deep time. McPhee pointed to the difficulties we have grasping it:

> Numbers do not seem to work well with regard to deep time. Any number above a couple of thousand years—fifty thousand, fifty million—will with nearly equal effect awe the imagination to the point of paralysis.[13]

McPhee may have been optimistic in implying that we can appreciate even a thousand years, given society's inability to comprehend climate change that has occurred over the past century and to grabble with the potential severity of more change in the coming decades.

Arguably, we can grasp in some sense the small scale of microbes more so than the immensity of deep time. At least we learned about the size of microbes long before we discovered the deepest of deep time on Earth: the age of the planet. In the seventeenth century when van Leeuwenhoek first peered at microbes with his primitive microscope (Chapter 1), it was thought that Earth was only a few thousand years old.[14] (In fact, it's about 4.5 billion years old.) In 1650–54, after thoroughly studying the available evidence in the Bible and secular records of comets, eclipses, and the discovery of stars, the Irish historian James Ussher concluded that Earth was created in the year 4004 BCE. It is easy to be amused by Young Earth ideas and to assume Ussher and others were blinded by their faith, but these men were erudite, thorough scholars who based their conclusions on the information available at the time. One such scholar was Isaac Newton. Along with revolutionizing physics and mathematics, Newton worked on Earth's chronology and proposed that Earth was created in 3988 BCE. The brilliance of these scholars couldn't compensate for their lack of understanding basic geology, among many deficiencies, and they didn't have the technology and the instrumentation needed to gather data relevant to Earth's origin. By contrast, seventeenth-century technology was far enough along for van Leeuwenhoek to invent a microscope powerful enough to see his animalcules, the "little animals." Leeuwenhoek made his discoveries about 300 years before the age of Earth was finally figured out in 1956.

Microbes seem far removed from the slow carbon cycle, yet we've already seen one microbial contribution: the calcium carbonate made by an alga that eventually formed chalk and the White Cliffs of Dover. Next, we'll see other ways microbes help make carbonate-rich rocks and build mountains.

From Microbes to Mountains

Coccolithophores and foraminifera in the open ocean are the main producers of calcium carbonate today, but their contribution to calcium carbonate production diminishes the further back in time we go. Pelagic calcifying organisms are latecomers to the carbon cycle. Although "naked" foraminifera without calcium carbonate shells arose as early as 1150 million years ago according to DNA sequence data, the fossil record indicates foraminifera with calcium carbonate didn't appear until the Early Cambrian about 500 million years ago.[15] The biggest contributor to calcium carbonate formation in today's ocean, coccolithophores, evolved even later, first appearing in the late Triassic period, and did not become prominent until around 93 million years ago.[16] These pelagic organisms provided calcium carbonate for one form of limestone, chalk, but they weren't around to make other forms.

Rather than by pelagic organisms, most limestone, even on the highest mountain, was formed by benthic organisms in shallow seas.[17] The shallow depths ensured bottom-dwelling communities received enough sunlight to support biological production, and seas had the high concentrations of calcium needed to make calcium carbonate. Although some of the oldest limestone may have been formed without biology's help, most is thought to have started as calcium carbonate in organisms or associated with organisms that make reefs. The foundation of a reef is made of cement—calcium carbonate. Several organisms, including oysters, mussels, and some polychaetes, make reefs in today's ocean, and colorful coralline algae that have calcium carbonate cell walls add to reef formations. But the most spectacular reefs are made by corals. And corals have made mountains.

The link between coral reefs and geology dates back to Charles Darwin's work on the formation of islands in the Pacific Ocean and to studies by Baron Ferdinand F. von Richthofen (1833–1905), who hypothesized that much of the Dolomite Mountains in Italy are ancient coral reefs.[18] Born seven years before Darwin published his coral reef work, Richthofen is better known, if he is known at all, for being the uncle to another Baron von Richthofen, the World War I flying ace Manfred von Richthofen ("The Red Baron"), and for his extensive work in China, including seven trips through most of the provinces of the Middle Kingdom.[19] He coined the phrase "silk road," or "Seidenstraße" in the original German. Although celebrated for establishing geography as an academic discipline in Germany, Richthofen was trained as a geologist and spent part of his time in China looking for coal deposits, after working for many years in the silver and gold mining areas of California and Nevada. His career as a geologist started in Europe.[20] In the summer of 1856, having just graduated from the University of Berlin, Richthofen accompanied several distinguished geologists on expeditions to the mountain ranges of Hungary, Transylvania, and Italy where he made the link between coral reefs and the Dolomites. He put forward his hypothesis in "Geologisohe Besobreibung der Umgegend von Predazzo," published in 1860. The idea that corals gave rise to these mountains of limestone wasn't accepted by geologists until nearly two decades later when a study published in 1879 demonstrated the reef-like nature of the Dolomites in South Tyrol.[21] Other geological features that started as reefs include the Capitan Formation in western Texas and southeastern New Mexico and similar formations in the high mountains of Nevada.[22]

Reefs and limestone-rich mountains owe their existence to microbes. From the corals that became the Dolomites during the Triassic to the Great Barrier Reef of today, reef-building corals depend on microbes.[23] These corals are a symbiosis between a marine invertebrate in the Cnidaria phylum, which also includes jellyfish and sea anemone, and a type of algae called dinoflagellates, housed in a structure called the coral polyp (Fig. 6.2). The dinoflagellate symbionts feed their animal host organic chemicals they make during photosynthesis, while in return the host releases nutrients like ammonium and phosphate needed by the dinoflagellate. The animal host also makes the calcium carbonate to protect itself and the symbiont from predation and viral attack. Because calcium carbonate formation in reef-building corals depends on photosynthesizing dinoflagellates for nearly all their food, big reefs are in shallow waters with lots of sunlight. There are exceptions. Stony corals calcify, even though some are too deep and too far away from sunlight to harbor photosynthetic symbionts (they feed on zooplankton passing by) and some

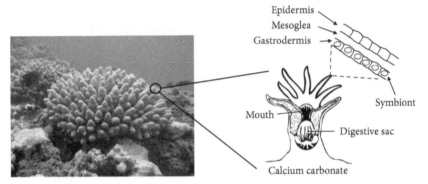

Epidermis
Mesoglea
Gastrodermis
Symbiont
Mouth
Digestive sac
Calcium carbonate

Figure 6.2 A hard coral (left) and one of its polyps (right), consisting of a coral invertebrate with a digestive sac and tentacles at its mouth extending into the surrounding seawater. The coral symbionts live in the inner layer of the tentacles. Each polyp is at most a couple of millimeters wide by a few millimeters to centimeters tall. The picture was taken by Andy Collins, U.S. National Oceanic and Atmospheric Administration.

soft corals have symbionts but don't have hard skeletons, only small calcareous structures called sclerites. Yet the biggest and the most spectacular reefs with the most calcium carbonate are built by corals assisted by photosynthetic dinoflagellates.

The dependency of corals on their microbial symbionts becomes all too clear during coral bleaching. More and more frequently, environmental conditions are so harsh that the coral invertebrate is forced to evict its colorful symbiotic algae, leaving behind the mostly colorless invertebrate and the white calcium carbonate.[24] When harsh conditions stress the algal symbiont enough, its photosynthetic apparatus produces peroxides, superoxides, and other toxic chemicals from oxygen, collectively known as reactive oxygen species. (Consider hydrogen peroxide's power as a disinfectant and as a hair lightener.) These oxygen-derived toxins are of more immediate danger to the coral than the prospect of going without the food provided by the symbiotic algae. Corals do not immediately die without their symbionts, and they can recover if conditions return to normal quickly enough. But corals are devastated when forced to forego their symbiotic dinoflagellates for long. Although pollution, disease, or abnormally cold water can induce coral bleaching, the most common cause is abnormally hot water.

Throughout their evolutionary history, corals and their algal symbionts have had to contend with warmer-than-average surface waters that occur during natural oscillations in climate like El Niño; reports of coral bleaching date back to the late nineteenth century.[25] What has changed is the frequency of marine heatwaves even in years without El Niño. Accompanying the upward trend in global temperatures, marine heatwaves have become more common and have lasted longer (Chapter 4). As a result, the time between excessively warm waters experienced by corals is now only about six years, much shorter than the 27 years corals had of respite after a heatwave hit in the early 1980s. Since 2000, about a third of bleaching events were followed by another within just one to three years. Even fast-growing coral species need 10 to 15 years to recover. Many coral reefs around the world are now in trouble because of global warming.

Symbiotic dinoflagellates offer a bit of hope for corals contending with marine heatwaves. Coral biologists have noticed that some corals more than others are susceptible to bleaching during a heatwave, and in experiments some corals survive in warm waters better than others. One key to the coral's survival is its symbiotic algae. Several different dinoflagellates that vary in their tolerance for high temperature are potential symbionts in corals.[26] The same coral may harbor more than one symbiotic dinoflagellate, and the symbiont that is most abundant within that coral can vary with water depth, light level, and other environmental conditions. To explore how the symbiont may help a coral survive a marine heatwave, coral biologists looked at how a species of stony coral, *Acropora millepora*, known to be susceptible to bleaching, would fare if exposed to warm water.[27] They transplanted 22 colonies of this coral from a reef in relatively cool waters to another reef further north along the northeast coast of Australia where surface waters were warmer. In the cool-water reef, the dominant symbiont in most of the coral was a type of dinoflagellate called at the time Clade C; all colonies of this coral also had smaller numbers of another type, Clade D. (Coral symbionts were once put into a single genus, *Symbiodinium*, and the variants, or clades within this genus, were labeled A, B, C, and so on. Recently these clades were upgraded to genus level and given full scientific names.[28]) When moved to the warmer waters, the corals bleached, and a few even died. But within a few months most regained their color and returned to good health, thanks to the revival of one of the symbionts, Clade D. The heat tolerance demonstrated by some symbionts may help corals survive the growing threat of global warming.

The Other Carbon Dioxide Problem

Unfortunately, corals and other calcifying marine organisms, including coccolithophores, are facing another climate-change problem caused by carbon dioxide increasing in the atmosphere: ocean acidification.[29] As oceans take up more and more carbon dioxide released by fossil fuel burning and other human activity, the acidity of seawater is increasing. That's the same as saying the average pH of the ocean is decreasing. (By definition, pH is the negative logarithm of acidity, or the concentration of hydrogen ions (H^+). When a solution is neutral, the concentration of H^+ is 10^{-7} moles per liter and pH = 7.0.) The equation describing how dissolved carbon dioxide leads to acidity is:

$$CO_2 + H_2O \rightarrow HCO_3^- + H^+$$

where CO_2, H_2O, and HCO_3^- are carbon dioxide, water, and bicarbonate, respectively. Addition of atmospheric carbon dioxide into the ocean pushes the equation to the right, leading to more hydrogen ions and lower pH. The average pH of the ocean has already decreased by about 0.1 pH units since the start of the Industrial Revolution. That may not sound like much, because the size of the decrease is hidden by pH being on a log scale, but it's huge; the 0.1 drop in pH is equivalent to about a 30 percent increase in acidity, something not seen in Earth's history over at least 300 million years.[30] An increase in acidity potentially stresses calcifying organisms as it reduces the amount of one of the two ingredients needed to make the calcium carbonate, carbonate (CO_3^{2-}), due its reaction with H^+ as described by the following equation:

$$CO_3^{2-} + H^+ \rightarrow HCO_3^-$$

With higher acidity and more H^+, the equation is pushed to the right and the concentration of carbonate decreases. Coral reefs are especially vulnerable to ocean acidification in part because today's corals produce a less stable form of calcium carbonate, aragonite. Due to both ocean acidification and global warming, along with pollution from coastal development, it is feared that by 2100 carbonate production by corals will not keep up with the loss of calcium carbonate, resulting in the eventual demise of coral reefs.[31] Unlike global warming, coral symbionts don't seem to help corals deal with ocean acidification, although the topic hasn't been looked at extensively.

The impact of decreasing pH on today's most prolific calcifying organisms, coccolithophores, is less clear. Higher acidity does reduce carbonate production by many but not all species of coccolithophores,[32] yet the future for even acid-sensitive algae looks brighter than for corals. Kai Lohbeck and colleagues examined whether the iconic coccolithophore *Emiliania huxleyi* could adapt to higher carbon dioxide levels and lower pH in an experiment that lasted for about a year.[33] That's not enough time for large organisms to evolve, but it is for these fast-growing algae. Lohbeck and colleagues subjected some *Emiliania* clones to high carbon dioxide levels for over 500 generations while they kept other clones at low levels as a control. They found that *Emiliania* could partially adapt in terms of both growth and calcification. Perhaps the alga could fully adapt if given more time. So even today's coccolithophores seem to be okay with more acidity. In two-week-long experiments with North Atlantic seawater, other investigators found that raising carbon dioxide levels and lowering pH induced a bloom of fast-growing coccolithophores with little calcium carbonate.[34] The alga apparently needs high carbon dioxide levels for rapid growth, perhaps because its carbon dioxide–fixing mechanism is even more inefficient than in other algae. (This inefficiency may be another reason why coccolithophores make coccoliths. Producing calcium carbonate during coccolith formation raises carbon dioxide levels within the coccolithophore.) The stimulation of coccolithophore growth by rising carbon dioxide helps to explain why these algae have become more common in the North Atlantic Ocean over the last 50 years.

Some models predict more coccolithophores in future oceans, even though ocean acidification and warming could reduce calcification by as much as 75 percent.[35] These algae could be winners with climate change while the oceanic carbon cycle loses. In the future, there may be less calcium carbonate not only for the carbonate pump but also for ballast that helps organic detritus sink to deep waters as part of the soft tissue biological pump (Chapter 4). Without being weighed down by calcium carbonate, organic particles would sink slower, decompose faster, and release more carbon dioxide before leaving the surface layer. Less carbon dioxide would be sequestered away from the atmosphere. The biological pump doesn't appear to be disrupted by climate change so far (Chapter 4), but it may be in the future if the ocean continues to warm and acidify.

Regardless of climate change, coccolithophores, foraminifera, and other calcifying microbes are huge in the carbon cycle. Along with corals and their microbial symbionts, these calcifying microbes have left behind the biggest stock of carbon on the planet, carbonate rocks. Next we'll discuss rocks made by another type of calcifying microbe. What's fascinating about these microbes and their rocky remains is what they have to say about the evolution of early life on Earth.

The Oldest Rocks and Early Life

The main calcifying organisms change once again as we go back further in geological time, to the earliest days of Earth. Not just coccolithophores and foraminifera but also corals weren't around when the oldest carbonate rocks formed billions of years ago. The oldest rocks are from the Precambrian, which started when the planet formed about 4.5 billion years ago and ended about 540 million years ago. The Precambrian covers over 85 percent of Earth's history. Not much is left from the Precambrian, its sedimentary rocks having been transmuted into metamorphic rocks by plate tectonics, destroyed by erosion and weathering, or buried beneath Phanerozoic debris. Rocks older than about 2.5 billion years only make up about 5 percent of the total record.[36] The origin of what has survived is difficult to decipher, and fingerprints of biology are hard to see. Abiotic mechanisms should be considered as likely as biotic ones in forming sedimentary rocks during the Precambrian. When biology was the mechanism, however, it was totally microbial, and Precambrian microbes were nearly entirely all prokaryotes: cyanobacteria, other bacteria, and archaea.

Paleontologists have gathered several pieces of evidence indicating the role of microbes in forming calcium carbonate rocks during the Precambrian. An example can be found on Spitsbergen, an island in the Svalbard archipelago about halfway between the northern coast of Norway and the North Pole. It is home to the Svalbard Global Seed Vault, the research station of Ny-Ålesund, and coalmines in Barentsburg and Pyramiden once owned and operated by the Soviet Union.[37] (A huge statue of Lenin still stands in Pyramiden.) With little soil and no forests or agriculture, the island's landscape is dominated by glaciers and geology, filled with red and brown sandstones and white carbonate rocks, such as those in the Akademikerbreen Group.[38] The carbonate rocks were not formed at this frigid latitude, north of the Arctic Circle. Rather, they were laid down along a coast in the tropics during the late Proterozoic and then were pushed north by tectonic forces, eventually arriving at their current location more than 100 million years ago. Rocks in the Akademikerbreen Group are mostly limestone, with laminated dolomite at the ancient shore's edge. (The mineral dolomite is calcium magnesium carbonate.) Fortunately for paleontologists, the Arctic environment has preserved signs of the rock's creators, microbes.

Some of the signs are readable only by paleontologists, while others are clear to everyone. The rocks have fragments of lipids known to be produced by bacteria, and the stable isotope composition of the rock's organic carbon and carbonate is consistent with a microbial origin. The ratio of the two nonradioactive or stable isotopes of carbon, ^{13}C and ^{12}C, is low in chemicals made by photosynthesis compared to those made by other processes (see Box 6.1 for more details). The $^{13}C/^{12}C$ ratio of organic carbon in these Spitsbergen rocks is characteristic of photosynthesis, suggesting cyanobacteria or other photosynthetic bacteria supported the ecosystem when these carbonate minerals formed in the late Proterozoic 630 to 540 million years ago. More evidence of bacteria at work comes from sulfur isotopes. Other evidence would convince even the nonspecialist that microbes were present when Spitsbergen rocks formed: paleobiologists found that Spitsbergen carbonate rocks house microfossils resembling filaments of modern cyanobacteria. The appearance of the Spitsbergen rocks is also suggestive of complex microbial communities. Viewed from a distance, the stack of limestone and related rocks is massive, soaring thousands of meters high, extending for over

Box 6.1 Microbial Fingerprints in Carbon

We've already seen the application of a radioactive, thus unstable carbon isotope, ^{14}C, for estimating the age of organic carbon in soils and the oceans. The amount of a stable carbon isotope, ^{13}C, relative to the most abundant carbon isotope, ^{12}C, is used to identify the reaction producing the carbon-containing chemical. Organisms treat the two isotopes nearly the same, except reactions are a bit slower with the heavier ^{13}C; how much slower depends on the reaction. Because photosynthesis prefers the lighter ^{12}C-carbon dioxide over the heavier ^{13}C-carbon dioxide, the organic chemicals made by algae and cyanobacteria have less ^{13}C and a characteristic low $^{13}C/^{12}C$ ratio. The $^{13}C/^{12}C$ ratio is a fingerprint left by microbes, implicating them in the formation of carbonate rocks on Spitsbergen and elsewhere.

a hundred kilometers. A closer inspection reveals the rock is constructed out of many layers, each a millimeter or so thick. Although lamination of this sort can be made by geochemistry without biology's help, it is reminiscent of microbial mats laden with calcium carbonate that grow today in a few locations around the world. The chemical data, fossil bacteria, and the structural similarities between these rock formations and known calcifying microbial mats all point to bacteria as the builders of carbonate-rich rocks in the Precambrian.

The way mat microbes make calcium carbonate is very different from how coccolithophores do it. The alga has dedicated membrane-enclosed pouches, or vesicles, where calcium-binding polysaccharides direct the formation of the calcium carbonate mineral calcite.[39] After enough calcite forms, the resulting coccolith is extruded to the outer surface of the coccolithophore. The entire process is an example of "biologically controlled biomineralization";[40] the alga governs all aspects of how the mineral is made, and it's clear the resulting mineral structure, the coccolith, is advantageous to the coccolithophore. In contrast, the way mat microbes make calcium carbonate is an example of "biologically induced biomineralization." Rather than close oversight and governance, mat microbes only indirectly promote conditions favorable for the formation of calcium carbonate. And the calcium carbonate conveys no obvious advantage to the microbes. Also, unlike coccolithophores, which make only calcite, microbial mats have other types of calcium carbonate, including aragonite and calcite with high amounts of magnesium.

Cyanobacteria are the keystone of calcifying microbial mats, although probably not of calcification. These photosynthetic microbes make organic chemicals that support the growth of other microbes in the mat and exude slimy extracellular polymers that hold the mat together and trap sediments and ions such as calcium while also protecting the cyanobacterium from dissection and harsh sunlight, two problems encountered on exposed tidal flats. Cyanobacteria seem prime candidates to contribute the most to making calcium carbonate because of how their photosynthetic activity affects the reaction for calcium carbonate formation (see the first equation in this chapter). By consuming carbon dioxide and increasing the pH, cyanobacteria pull the reaction to the right, leading to calcium carbonate formation. However, the work of cyanobacteria in making calcium carbonate is undone at night when photosynthesis stops in the dark and heterotrophy takes over, at which point the degradation of organic chemicals begins. Heterotrophy produces carbon dioxide and drives pH down, leading to the dissolution of calcium carbonate. The calcium and bicarbonate ions go back into solution. So cyanobacteria aren't responsible for the calcium carbonate. If they aren't, what is?

Sulfate-reducing bacteria are likely suspects. These bacteria are abundant and active deep in the mat where calcium carbonate accumulates and dissolved oxygen is gone. A few millimeters below the mat's surface, dissolved oxygen disappears as heterotrophic bacteria and others use it more quickly than it can be replenished by diffusion. When oxygen is depleted, sulfate-reducing bacteria take over and calcium carbonate builds up.[41] These bacteria use sulfate instead of dissolved oxygen to degrade organic chemicals, and they release hydrogen sulfide as a byproduct. Sulfate-reducing bacteria also degrade cyanobacterial extracellular polymers that tie up calcium ions and prevent them from joining with bicarbonate to form calcium carbonate. By raising the pH and freeing calcium ions, sulfate reducers promote calcium carbonate formation. Sulfate-reducing bacteria also appear to have a special role in making a type of calcium carbonate rock, dolomite, which is second only to limestone among sedimentary rocks. Sulfate reducers are one solution to the "dolomite problem"; even though chemical conditions seem ripe for formation of this carbonate mineral at a normal temperature and pressure, it has not been observed in the absence of biology. Dolomite does form, however, in the presence of sulfate-reducing bacteria. When C. Vasconcelos and colleagues added some "black sludge" from Lagoa Vermelha, a coastal lagoon near Rio de Janeiro, Brazil, to an incubation with seawater salts and simple organic chemicals, they saw the production of hydrogen sulfide, which is indicative of sulfate-reducing bacteria, and the formation of dolomite.[42] Other studies have suggested dolomite is made by sulfide-oxidizing bacteria[43] that live on the sulfide produced by sulfate-reducing bacteria, and still other studies have connected methane-producing archaea to dolomite. How these microbes make dolomite is still not entirely clear, nor is it known if they are responsible for making all of it. Regardless, the work indicates microbes can promote the formation of dolomite and other carbonate-rich rocks.

Microbial mats are everywhere today, but one type, stromatolites, is rare. A stromatolite is a microbial mat with distinctive layers of cyanobacteria, other bacteria, and calcium carbonate. One layer builds on another as the mat traps sediment and calcium carbonate accumulates. The most famous location with modern stromatolites is Hamelin Pool on the western end of Australia (Fig. 6.3), where stromatolites cover about 1400 square kilometers along a 135-kilometer shoreline.[44] Two other places are the Exuma Cays in

Figure 6.3 Modern stromatolites in Hamelin Pool, Western Australia. Each stromatolite is about 0.5 meters in diameter. Photograph taken by Paul Harrison. CC BY-SA 2.0

the Bahamas and Baja California.[45] Even though they can be big in a few locations, modern mats are nothing compared to those in the geological past.

The heyday of stromatolites and other carbonate-rich microbial mats was in the Precambrian over 500 million years ago. The Akademikerbreen Group on Spitsbergen is a monument to the time when microbial mats regulated much of the carbon cycle on the planet. They were the only reefs around back in the Precambrian, analogous to the Great Barrier Reef of today, but more massive, some soaring hundreds of meters high and stretching for hundreds of kilometers along ancient shorelines. They declined at the end of the Precambrian, perhaps due to the evolution of primitive animals that could browse or burrow into mats or outcompete mat-making microbes for space. One successive competitor was reef-building corals, which evolved in the early Ordovician, 485 to 445 million years ago.[46] The few modern habitats with extensive living stromatolites have high salinity or high alkalinity (high pH) or both—conditions that exclude animals and other reef builders. The salinity of Hamelin Pool with its massive living stromatolites is about twice the level of normal seawater. However, the decline in stromatolites doesn't perfectly align with the rise of animals, leaving room for other explanations.[47] Regardless of why they declined, stromatolites are testimonies to the age when microbes ruled the planet even more so than they do today.

Ancient stromatolites are more than just curios from the distant geological past. Stromatolites have provided candidates for the oldest fossils and the first signs of life on Earth. The first evidence for Precambrian life was discovered in 1953 by Stanley Tyler and Elso Barghoorn in the Gunflint Formation, a nearly 2-billion-year-old stromatolite in northern Ontario.[48] Tyler and Barghoorn made their discovery only a couple of years before the Australian geologist Philip Playford found stromatolites that were very much alive in Hamelin Pool, providing strong evidence that ancient stromatolites could be produced by biology. Currently, the oldest fossils of microbes recognized by many paleontologists are in northwestern Australia embedded in stromatolites almost 3.5 billion years old. More than establishing when life began, stromatolites have provided clues about where it started. The 3.5-billion-year-old stromatolites in northwestern Australia, for example, have been used to argue that life began in a hot spring on land rather than in the hot water of a hydrothermal vent in the deep sea.[49] Either location would fit DNA sequence data and other evidence suggesting that the first cell may have been similar to today's hyperthermophilic bacteria that grow best in water hotter than 80 degrees Celsius.[50] Fossil stromatolites are likely to continue to provide answers to questions about when and where life first appeared on Earth.

Returning to the carbon cycle, stromatolites and other microbial mats add to the list of ways microbes make carbonate rocks. How mat microbes promote the formation of calcium carbonate is quite different from the controlled mechanism used by coccolithophores, foraminifera, and the dinoflagellate-coral symbiosis, yet the end result is similar. All contribute to the largest carbon stock on the planet. As pointed out in Chapter 2, carbon in sedimentary rocks, about 150,000,000 petagrams, dwarfs all other carbon stocks, including the 1328 petagrams in fossil fuels and the 870 petagrams now in the atmosphere. Recall that two other large stocks of carbon, refractory organic carbon in soils and in the ocean, also are formed with assistance from microbes (Chapters 3 and 4). Our climate would be very different if all that carbon were in the atmosphere as carbon dioxide.

However, carbonate rocks are not a dead end, a place where carbon dioxide is sequestered away forever. The carbon in rocks is eventually recycled as part of the slow carbon cycle. Just as the making of carbonate rocks can involve microbes, so too can the recycling of calcium carbonate back to its starting ions.

The Weathering Thermostat

At first glance, geology doesn't seem to need any help from microbes to break up calcium carbonate and return it to its dissolved constituents. Calcium carbonate accumulating in deep ocean sediments over countless years is eventually moved by plate tectonics until it is sucked into Earth's mantle at a subduction zone, where one tectonic plate slides under another.[51] Once in the mantle, the calcium carbonate undergoes metamorphosis, releasing carbon dioxide into the atmosphere via volcanic eruptions. The entire process takes millions of years, maybe even hundreds of millions of years: the domain of geology, not microbiology. Another way to recycle carbonate, however, can involve microbes.

Calcium carbonate deposited in shallow seas can be uplifted and exposed to air during episodes of mountain building. The newly exposed rocks undergo "weathering," the term geologists use to describe the breakdown of rocks and minerals. Now microbes come into the picture. Before getting to the microbiology, however, we need to go over more geochemistry. Key to the two chemical weathering reactions to be discussed here is carbonic acid (H_2CO_3) formed when atmospheric carbon dioxide dissolves in water (Fig. 6.4). Carbonic acid attacks calcium carbonate and releases one calcium ion and two carbonate ions. The reaction also consumes one molecule of atmospheric carbon dioxide, which cancels out the carbon dioxide molecule released during calcium carbonate formation. Calcium carbonate

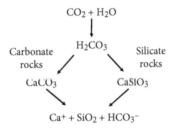

Figure 6.4 Chemical weathering of carbonate and silicate rocks. The reaction of carbonic acid (H_2CO_3) with calcium carbonate ($CaCO_3$) consumes one carbon dioxide molecule (CO_2), and this balances out over millions of years the one molecule of carbon dioxide released during calcium carbonate formation. The reaction with silicate rocks ($CaSiO_3$) consumes two carbon dioxide molecules. Unlike carbonate rocks, formation of silicate rocks doesn't release any carbon dioxide.

formation and weathering are thought to be in balance over a million-year time frame, although perhaps not over 100 to 10,000 years.[52] Another weathering pathway that consumes carbon dioxide is the reaction of carbonic acid with silicate minerals, such as $CaSiO_3$, or wollastonite; most silicate rocks have other elements, such as aluminum and sodium, and more complicated chemical formulas. Unlike calcium carbonate rocks, silicate rocks participate in the slow carbon cycle without the cost of releasing carbon dioxide during their formation.

The consumption of carbon dioxide by weathering is key to regulating levels of atmospheric carbon dioxide and to setting Earth's global temperature over millions of years.[53] It's a global thermostat. An increase in carbon dioxide and temperature pushes weathering reactions to speed up and consume more carbon dioxide. Warmer temperatures also accelerate the hydrological cycle and thus weathering, because warmer air holds more water which yields more rain and higher runoff, exposing rocks to more water and weathering. We can see this today in the tropics where runoff is much greater than in cooler latitudes, and weathering is faster. If the planet cools too much, weathering diminishes as the hydrological cycle slows down, allowing atmospheric carbon dioxide and global temperatures to rebound. The time scale of weathering for regulating Earth's climate is incredibly long, perhaps hundreds of thousands of years. It operates in deep time. So it's not the answer to the current buildup of atmospheric carbon dioxide that we've caused as weathering will not reduce carbon dioxide released by human activity soon enough to stop the climate change now occurring. Yet the weathering thermostat has been powerful enough to keep Earth's climate from going to extremes for the last 600 million years, at least. Although both global temperature and carbon dioxide levels have varied with the coming and going of ice ages, there has been no long-term trend up or down. Weathering is too slow to stop all temperature swings, but eventually it does restore the planet's climate to its long-term average.

Weathering of carbonate and silicate rocks can be amplified by microbes. Once a rock is exposed to air or water, it is rapidly colonized by cyanobacteria, other bacteria, fungi, algae, and lichens (a symbiosis between fungi and an alga or cyanobacterium), setting in motion forces that end in the rock's eventual demise.[54] After being ensconced in cracks and crevices, microbes start to work away on the rock and its minerals. Respiration by heterotrophic bacteria and fungi increases acidity that damages the rock, while acidic extracellular polymers used by microbes for protection have the same effect. The retention of water by microbial polymers of all sorts enhances weathering reactions at the rock surface, although microbial polymers can also blanket rocks too effectively and hinder weathering. Some microbes excrete organic acids like oxalate to extract phosphorus nutrients bound in rocks, and other microbes release specific organic chemicals, called siderophores, to extract iron and other metals needed for metabolism. Some bacteria contribute to breaking up rocks by oxidizing iron and sulfur in minerals such as pyrite, better known as fool's gold, to gain energy. Other bacteria oxidize manganese in rock minerals, though it's not clear why they do so. Even if carbon dioxide isn't directly involved, these extractions and oxidations help to break up rocks and expose more surface area for carbon dioxide–consuming reactions. In short, extractions, oxidation reactions, acidification, and water retention carried out by microbes all contribute to weathering and the sequestration of atmospheric carbon dioxide.

The fingerprints left behind by the microbial assault on rocks are microscopic, yet the cumulative result can be macroscopic. Scanning electron microscopy reveals telltale pitting and etching of rock surfaces, some accompanied by bacterial cells or fungal filaments (hyphae), which take advantage of imperfections and bore into seemingly impenetrable rocks. A microscope isn't needed to see other microbial handiwork. The Lower Kane Cave in Wyoming was carved out by the sulfuric acid produced by sulfur-oxidizing bacteria,[55] while intertidal carbonate rocks along the coast of Britain have been weakened by cyanobacteria.[56] Hydroxyl ions released during photosynthesis by cyanobacteria have sculptured sandstone mountains and outcrops in South Africa.[57] Microbes aren't confined to natural stones and rocks. Built structures such as monuments and buildings can deteriorate because of microbes.[58] Erosion aided by microbes over 300 years has rendered inscriptions on tombstones in the United Kingdom illegible, and a microbial film discolored the Vermont marble of the Jefferson Memorial in Washington, DC, so badly that its restoration cost the National Park Service $14.5 million in 2021.[59] An emerging field of science, biogeomorphology, explores how microbes and other organisms shape the geological landscape.

More than just disfiguring tombstones and monuments, weathering aided by microbes has radically altered Earth's climate in the geological past. Intense silicate weathering is one hypothesis to explain the onset of several early ice ages that occurred between 1000 and 290 million years ago.[60] The Carboniferous ice age around 300 million years ago was set up by the carbon dioxide drawn down during the late Devonian about 400 million years ago when trees evolved and lush forests appeared on land. Along with contributing to the formation of soils and the sequestration of atmospheric carbon dioxide in soil organic matter, trees sped up weathering by promoting the release of carbon dioxide below the soil surface where it could build up to levels much higher than seen in the atmosphere. The carbon dioxide came from respiration by tree roots and by heterotrophic bacteria and fungi degrading organic chemicals released by roots (Chapter 3). Going back further in geological time before vascular plants covered land, only microbes were around to potentially enhance weathering. Experiments have demonstrated that bacteria can speed up the breaking apart of some silicate rocks, such as feldspar, one of the most common silicate rocks on Earth, although microbes are powerless against other silicate rocks.[61] Lichens may have been especially important in boosting weathering on early continents without lush vegetation, considering what lichens are capable of doing today. Compared to bare rock, lichens accelerate by over 20 times the weathering of gabbro boulders in the Norwegian Arctic[62] and of lava on the Big Island of Hawai'i.[63] Whether bacteria, fungi, or the microbial symbionts that make up lichens, microbes likely helped weathering to consume atmospheric carbon dioxide and to trigger several ice ages on Earth millions of years ago.

The contribution of microbes to weathering is like the other connections microbes have with the slow carbon cycle. These connections operate at the scale of a few microns—smaller than a human hair is thick. It would amount to nothing if microbes were at work for only a few days or even several years. Yet by chipping away relentlessly for eons and eons, the incremental action of microbes adds up. The White Cliffs of Dover were formed over millions of years one micron-thick coccolith at a time. By building mountains while also wearing down rocks, microbes are the unseen agents that have helped the slow carbon cycle regulate Earth's climate over millions of years.

7

Natural Gas

We've focused on carbon dioxide so far for good reason. It's the biggest greenhouse gas, accounting for nearly 70 percent of global warming caused by humans (Chapter 1). Yet the second biggest greenhouse gas, methane, has been drawing a lot of attention recently, one sign being the Global Methane Pledge in which the United States, the European Union, and over a hundred other countries vowed to reduce methane emissions 30 percent by 2030.[1] Methane has been singled out because it is a potent greenhouse gas. A molecule of methane has 80 times more impact than a carbon dioxide molecule does over 20 years and nearly 30 times over 100 years, the time frame used by the Intergovernmental Panel on Climate Change (IPCC) for its calculations.[2] Because its lifespan in the atmosphere is short, only about a decade, reducing methane emissions would quickly translate into drops in atmospheric levels and less global warming. Lower methane levels would also improve air quality by decreasing the formation of ozone in air near land surface. While ozone in the stratosphere is needed to minimize DNA-damaging ultraviolet light, ozone close to farms and where people live reduces crop productivity and causes respiratory diseases. Among efforts to reduce greenhouse gases, methane is said to be low-hanging fruit. The technology for substantially cutting methane emissions is already available and is relatively cheap, and there is an economic incentive to reduce methane losses and capture it before it leaks into the atmosphere, given that it's the main gas in natural gas.

Despite being low-hanging fruit, efforts to reduce emissions have been ineffectual to date, and methane has been increasing in the atmosphere. Bubbles trapped in ice at Law Dome in Antarctica have provided data about methane and its sources for the centuries before the Industrial Revolution,[3] and continuous monitoring at about 50 sites worldwide, spearheaded by the U.S. National Oceanic and Atmospheric Administration (NOAA), has documented in more detail atmospheric levels since 1983.[4] The bubbles and the monitoring sites indicate that atmospheric methane has increased nearly three-fold over the preindustrial average, much more than the 50 percent

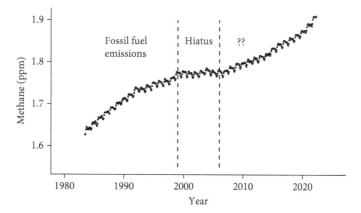

Figure 7.1 Methane levels in the atmosphere since 1983. Concentrations are in parts per million (ppm). The rise in methane from 1983 to 1999 was due to fossil fuels. It's not entirely clear why methane began to increase again after the hiatus ended in 2006. Data from Ed Dlugokencky, NOAA.

increase seen for carbon dioxide. But the increase over the last 40 years hasn't been smooth (Fig. 7.1). Methane did march straight up once monitoring started in 1983, but then net emissions unexpectedly stalled between 1999 and 2006 when levels remained steady. Unfortunately, methane did not stay steady for long. Climate scientists were surprised and dismayed to see that after 2006, atmospheric methane resumed its upward trend, undeterred by the COVID-19 pandemic when fossil fuel production and use and presumably methane leaks declined. Methane levels and emissions are now at record levels.

The rise in atmospheric methane can only be partially tied to fossil fuels. The Antarctic ice core data suggest that methane rose slightly above natural levels during preindustrial years when forests and grasslands were cleared and wood and grass were burned. The much larger increase during the twentieth century was due to the growth of the natural gas industry but also to the extraction and use of other fossil fuels. Coal mining is a large source, accounting for a third of all methane emissions connected to fossil fuels according to the latest Global Methane Budget.[5] Along with the methane released from coal seams during mining operations, the gas also comes from coal waste piles and abandoned mines. Lower emissions from coal mining and other fossil fuel industries are one explanation for the static atmospheric methane levels seen from 1999 to 2006.[6] Why levels started to increase again in 2007 is not entirely clear, nor is it completely understood why levels have climbed even more swiftly in the last few years.

One explanation is that degradation of methane has slowed. Atmospheric methane is degraded mainly by purely chemical reactions with little help from microbes. Most degradation, 90 percent of the total, occurs when methane reacts with the hydroxyl radical (OH) in the atmosphere.[7] Sometimes called the detergent of the atmosphere, the hydroxyl radical also reacts with pollutants such as carbon monoxide, nitrogen and sulfur oxides, ozone, and hydrochlorofluorocarbons.[8] Because the hydroxyl radical has a lifetime of less than a second, direct measurements of its concentrations reflect only local conditions, forcing atmospheric chemists to use models to make global estimates. These models indicate that atmospheric levels of hydroxyl radical have varied from year to year, but there isn't much evidence that levels have diminished enough to explain the big increase in methane levels over the last 15 years. It does appear that less degradation by the hydroxyl radical partially explains why emissions continued to be high in 2020 despite lower methane emissions by human activity during the COVID-19 pandemic;[9] lockdowns and other restrictions imposed during the pandemic also reduced air pollutants like nitrogen oxides and ozone that produce the hydroxyl radical. However, for the other years, lower degradation is not the answer. One recent study estimated that only 14 percent of the recent rise in methane was due to a decrease in the hydroxyl radical.[10]

It will probably not be a surprise to learn that atmospheric chemists believe the big increase in atmospheric methane is due to higher emissions. What may be surprising is the source of those emissions, especially during the last few years.

A powerful tool for figuring out where methane comes from is its stable carbon content. As mentioned in Chapter 6, the amount of the rare ^{13}C isotope relative to the abundant ^{12}C isotope in a chemical gives a clue about which reactions produced or consumed that chemical (see Box 6.1; Box 7.1 gives a few more details about how ^{13}C data are expressed). What makes carbon isotopes so powerful is that there is a big difference in the ^{13}C content of methane produced by different mechanisms. Methane made by methanogens, what geochemists call biogenic methane, has a $\delta^{13}C$ of about −60 ‰. It has a lot less ^{13}C than the standard because methanogens discriminate against ^{13}C and produce ^{12}C-rich methane. Much of fossil fuel methane, the gas made by geological processes, what geochemists call thermogenic methane, has more ^{13}C and a larger (less negative) $\delta^{13}C$ of about −40 ‰.

Box 7.1 ^{13}C Notation

The carbon stable isotope ^{13}C provides many hints about methane sources. Rather than looking directly at the ratio of ^{13}C to the abundant carbon isotope ^{12}C in a sample, isotope geochemists compare the ratio in a sample to the ratio in a standard made with Pee Dee Belemnite, a limestone from South Carolina. The resulting number is preceded by the Greek letter delta (δ), commonly used in science to denote a difference between two numbers, and is then followed by ‰, or per mil (per one thousand), analogous to the percent symbol for percent (per one hundred). In geochemistry, relative ^{13}C levels vary from -12 ‰ found in the organic carbon from some plants to -60 ‰ or lower (more negative) for biogenic methane. Although the difference may seem small (much less than a percent), it is easily detected by a modern mass spectrometer and is huge to isotope geochemists.

There is even more ^{13}C in terrestrial plants; δ^{13}C is -25 ‰ for vegetation like trees and -12 ‰ for grass-like vegetation.

Changes in δ^{13}C convinced geochemists that the soaring methane wasn't coming only from fossil fuels. After holding steady from 1999 to 2006, δ^{13}C of methane started to drop substantially in 2007 just as methane concentrations resumed their rise.[11] The δ^{13}C of fossil fuel methane isn't negative enough to explain the drop. Although fossil fuel production has grown substantially over the years, methane emission as a fraction of fossil fuel production has declined, thanks to improvements in detecting and reducing methane leaks. Emissions from coal mining in China, today the country that emits the most methane, may have declined since 2012.[12] The drop in δ^{13}C also couldn't be due to the methane produced during forest fires or the burning of other plant material. That methane has more ^{13}C, not less. If not fossil fuels or burning plants, that leaves biogenic methane. But which biogenic methane? It's all produced by methanogenic archaea, yet those microbes live in many habitats, both natural and places quite far from nature. So which habitats?

One "habitat" is on farms. Agriculture is the largest source or the second largest source of methane on the planet, depending on how the methane budget is compiled (Fig. 7.2), and is one possible explanation for the recent rise. Emissions from agriculture were as much as 217 teragrams

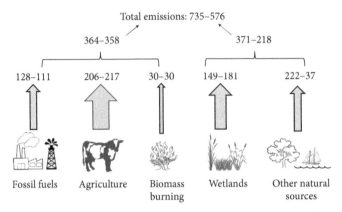

Figure 7.2 Methane emissions, in teragrams per year. The first number in the range is from the bottom-up approach for estimating emission rates and the second is from the top-down approach. "Biomass burning" refers to forest fires, the burning of wood and other plant material, and biofuels. Data are averages for 2008–2017 from the 2000–2017 Global Methane Budget.

(10^{12} grams) per year in 2008–2017, the latest period with good global estimates, nearly double the amount released by fossil fuel production and use.[13] Most of that methane came from cattle, sheep, and goats—ruminants that release methane by belches and flatulence. Manure from ruminants and other livestock is another source, and rice cultivation chips in more, another 30 teragrams annually. All this methane comes from methanogens. However, despite being a huge source of methane, agriculture doesn't seem to be the complete answer to why the gas has been increasing so much recently. The rise in cattle and rice production in various parts of the world didn't coincide with changes in methane levels over the last decade.[14] The culprit may be nature, perturbed by humans.

Inflammable Air and Marsh Gas

In a letter dated April 10, 1774, Benjamin Franklin wrote to the prominent English chemist Joseph Priestley about some "American experiments," with gases emitted by local waters that could burst into flames.[15] It was the year Priestley would isolate oxygen (what he called dephlogisticated air), one of many gases that he had discovered or co-discovered during the early days of chemistry. Franklin's letter to Priestley began by recounting what he had learned on a trip through New Jersey in 1764; a lit candle held over some of the area's rivers could start a fire that quickly spread over the water and burned for half a minute. A friend of Franklin learned more by stirring mud with his walking stick and applying a candle to the small bubbles that rose in the shallow waters. The bubbles also caught fire. Franklin thought the gas may be a "volatile oil of turpentine" from a nearby pine-filled swamp. Several years later, in the evening of November 5, 1783, only a few months after the American Revolution had officially ended, Thomas Paine reported similar results when General George Washington put a flame to the "impure and often inflammable air" that arose from mud of a river again in New Jersey.[16] (A video available on the Internet dramatically shows what happens when this inflammable air is ignited.[17]) The inflammable air had already been investigated more rigorously in 1776 when Alessandro Volta, after reading Franklin's letter, analyzed the gas bubbling from the mud of marshes near Lake Maggiore, Italy.[18] Volta described what he called *aria infiammabile nativa delle paludi* (inflammable air native to marshes) as differing from inflammable airs, or gases he could produce by other means. Along with his fame for work in electrochemistry, Volta is credited for discovering what we now call methane.

It was no accident that Franklin's friend and Volta observed methane in wetlands like the swamps and marshes in New Jersey and northern Italy. Poking around in bogs, sloughs, fens, muskegs, and mires also releases bubbles of the gas. Depending on how estimates are made, wetlands are the largest or one of the largest natural sources of methane on the planet (Fig. 7.2). They are the largest source, releasing 181 teragrams of methane each year, if the methane budget is compiled by the top-down approach, which starts with global atmospheric levels of methane and works down. Wetlands release less methane, 149 teragrams per year, according to estimates made by the bottom-up approach, which extrapolates from local observations to a global number. To put those numbers in perspective, wetlands and other natural sources release almost two to nearly three times more methane than fossil fuel industries do, according to the bottom-up and top-down approaches, respectively. The difference between the two ways of compiling methane emission rates is a sign of unknowns and of the difficulties in coming up with numbers for the entire planet, and many of these estimates may be revised with more research. But that doesn't change the fact that wetlands are a large source of methane.

After reviewing the carbon budget presented in Chapter 2, it may be surprising to see the ocean is not a big source of atmospheric methane, unlike carbon dioxide. Oceanic waters not only take up a lot of carbon dioxide from the atmosphere, they also produce a lot mostly because of microbes. Respiration by marine microbes releases about the same amount of carbon dioxide as do terrestrial plants and soil microbes. The ocean is a big part of every other biogeochemical budget if only because of its size. It is true that the oceans release some methane, explaining why surface waters are supersaturated with the gas; concentrations are higher than expected from the diffusion of atmospheric methane into seawater.[19] Some of this methane is from organisms other than the main producer of methane, methanogenic archaea, but these alternative producers don't make much methane on a global scale. Understanding why the oceans don't release a lot of methane is useful for understanding methane emissions from all environments.

The oceans don't produce much methane because most oceanic regions have high levels of oxygen and sulfate. Inhibition of methane production by oxygen is easier to understand. Methanogenic archaea are strict anaerobes, meaning that even a whiff of oxygen inactivates if it doesn't kill them. Despite ocean deoxygenation (Chapter 4), most of the ocean has levels of oxygen way too high for methanogens to survive and prosper. Oxygen does disappear eventually in a few isolated oceanic basins and in organic-rich coastal sediments, but then methanogens are hampered by high concentrations of the second chemical, sulfate. (Sulfate is third only to sodium and chloride among the major ions in seawater.) Sulfate has to decrease substantially before methane production can start. The negative effect of sulfate on methanogens is indirect; methanogens must compete against sulfate-reducing bacteria for the chemicals used by both as sources of energy. Without oxygen but with adequate sulfate, sulfate-reducing bacteria flourish, taking advantage of many energy sources, including organic matter and energy-rich inorganic chemicals. Methanogens on the other hand are fastidious and can use only a few chemicals like acetate and hydrogen gas. Unfortunately for methanogens, those few chemicals are taken up much more effectively by sulfate-reducing bacteria. As a result, when sulfate reduction is active, the concentrations of the few chemicals used by methanogens drop to levels too low to fuel much methane production. The inhibition of methanogenesis by oxygen and sulfate is one reason why oceans don't appear in diagrams like Figure 7.2.

Freshwater wetlands are a big part of those diagrams because they don't have much oxygen or sulfate, but they do have lots of organic material from wetland plant production, which is prodigious. Degradation of that organic material consumes all oxygen in wetland sediments faster than it can be replenished by diffusion or mixing with oxygen-rich water. Soils that become waterlogged also run out of oxygen and can emit methane. Soils, freshwater wetlands, and other freshwater habitats have little sulfate, the other big problem for methanogens in the oceans. Without sulfate, sulfate reduction is negligible and methanogens can thrive.

Wetlands were always high on the list of suspects for why methane has been rising so much over the last 15 years. Wetland methane with its low ^{13}C content explains the $\delta^{13}C$ data. More support for the wetland hypothesis came from data about how methane emissions vary around the world according to a network of stations that continuously monitor atmospheric methane and its $\delta^{13}C$. Data from these stations point to changes in the tropics as causing the recent rise in atmospheric methane.[20] The tropical wetland hypothesis has been supported by data gathered by another way to measure methane.

Methods for detecting methane range from the simple and crude to the sophisticated and precise. A leak from a natural-gas stove is obvious from the foul rotten-egg odor emanating from the sulfur chemical mercaptan added to natural gas; without it, methane is odorless. The leak perhaps could be found by using Benjamin Franklin's friend's approach and seeing where a lit candle ignites a flame, although it would be safer to coat suspected pipes with soapy water and look for bubbles. The gold standard, the approach used at the worldwide monitoring stations, is to collect air in glass flasks and to measure its methane content by gas chromatography with flame ionization detection. But that's too laborious for routine use and for sampling large areas. Several companies now market methane-detection instruments which range from handheld devices to those mounted on satellites. One study used data from an airborne infrared absorbance device operated by Kairos Aerospace was used to detect leaks during oil and natural gas operations in the New Mexico Permian Basin.[21] Similar instrumentation has been deployed on the Japanese Greenhouse Gases Observing satellite launched in 2009 and the European Tropospheric Monitoring Instrument on board another satellite launched in 2017. The European satellite was used to quantify the methane emitted during a natural gas well blowout in Ohio that lasted for 20 days.[22]

A group of atmospheric chemists led by Paul Palmer at the University of Edinburgh used data from the Japanese and European satellites to deduce methane sources worldwide and to explain why emissions have been increasing recently.[23] They concluded that 80 percent of the changes in atmospheric methane from 2010 to 2019 was due to emissions from the tropics, mostly Africa, followed by South America and India. Other studies have been even more explicit in saying those emissions came from tropical wetlands.[24]

What happened in those wetlands? Why did they start to emit more methane after 2010—and probably are continuing to do so today? The short answer is global warming. Along with warmer temperatures speeding up methane production by methanogens, atmospheric chemists have focused on changes in the hydrological cycle, likely intensified by global warming. The Palmer-led study linked higher methane emissions in east Africa and South America from 2010 to 2019 to changes in sea surface temperatures of the tropical Atlantic and Pacific Oceans caused by El Niño.[25] With El Niño came higher rainfalls and more methane from east Africa and South America while during the same years, both rainfall and methane emissions from Southeast Asia were lower. More rain falling on wetlands could boost methane emissions in at least two ways. With more water, wetlands and oxygen-free habitats optimal for methanogenesis expand. The additional water could also stimulate more plant growth and release of organic material that fuels methane production. Plant yields improved by rain could also help ruminant production in the tropics, another source of the extra methane that can't be discounted. More methane emitted by ruminants in the tropics would fit much of the data, especially the ^{13}C numbers. Methane in both wetlands and ruminants comes from methanogens. As Euan Nisbet put it, "A cow is a walking wetland."[26]

Except for raising more walking wetlands, human activity may not seem responsible for the most recent rise in methane, assuming the tropical wetland hypothesis is correct. It is true that rainfall intensity and frequency naturally vary with El Niño, more formally known as the El Niño-Southern Oscillation, and with the North Atlantic Oscillation, the Pacific Decadal Oscillation, and the Atlantic Multidecadal Oscillation. All these oscillations and the changes in rainfall they cause are natural. However, on top of this natural variation in precipitation and temperature are the changes we have brought on by global warming. Compared to the past, precipitation is already heavier in wet regions of the world and lighter in dry ones, a difference that is predicted to become starker with continuing climate change.[27] It's a classic case of positive feedback; more precipitation and higher temperatures lead to more methane and more climate change, causing more precipitation, higher temperatures, more methane, and so on. So, even if wetlands explain why methane emissions have been increasing recently, the real culprit is global warming caused by human activity. (Chapter 9 will discuss whether one way to reduce methane emissions is to fill in wetlands. The short answer is, no.)

Nearly all the discussion about rising atmospheric methane centers on the gas coming from fossil fuels or from methanogenic archaea in ruminants, rice fields, and wetlands. What's not highlighted enough by atmospheric chemists, however, is the importance of other microbes in degrading methane. There would be even more methane in the atmosphere if not for methane-degrading bacteria and archaea.

Degradation and Unlikely Symbioses

Atmospheric chemists always say that methane degradation is nearly entirely done by atmospheric chemistry. Along with the 90 percent of total degradation carried out by the hydroxyl radical, other chemical reactions in the atmosphere chip in another 5 percent.[28] The remaining 5 percent is attributed to soil microbes. All true—if we're talking only about methane in the atmosphere. However, much of the methane produced by methanogens or released from deep geological reserves is degraded before it ever reaches the atmosphere. As much as 80 percent of the methane produced in deep sediments in lakes is degraded in the water column before reaching the surface,[29] and more than 90 percent of the methane produced in rice fields is consumed by methane-degrading microbes, or methanotrophs.[30] Degradation is equally important in the natural environment that releases the most methane: wetlands. On average, microbes degrade between 40 and 70 percent of the methane produced in wetlands worldwide.[31] One study conducted near Utqiaġvik (Barrow), Alaska, found that methane degradation matched or even exceeded methane production in incubations with peat from a nearby tundra.[32] Methane degradation in the average wetland isn't as efficient as seen for the tundra peat, and it doesn't stop wetlands from releasing a lot on a global scale. Still, emissions would be even higher without methane degradation by methanotrophs.

That a lot of methane escapes degradation is probably not the fault of the methanotrophs. The ^{13}C of atmospheric methane suggests that most of the methane emitted by environments like wetlands doesn't encounter a methanotroph. Because methanotrophs favor ^{12}C-methane over ^{13}C-methane, any methane surviving methanotrophy would have more ^{13}C over time: the opposite from what has been observed. Methane escaping natural environments has low, not high relative amounts of ^{13}C, suggesting that it mostly bypasses methanotrophs. Two bypass routes are known (Fig. 7.3). One is bubbles, what biogeochemists call ebullition, percolating from mud or sediment, rising through the overlying water until they burst at the surface and release gases into the atmosphere. The methane in these bubbles can't be degraded by microbes unless it diffuses out into the water. That happens too slowly for much degradation unless the bubble forms in deep water. The second route bypassing methanotrophs is via plants. As atmospheric oxygen is transported by a wetland plant down to its roots to fuel root metabolism, methane goes up in the opposite direction via diffusion. Being a small, uncharged molecule, methane easily crosses membranes by diffusion, going from high levels around roots to low levels inside the plant vascular system. Once inside the plant, methane then can continue to the surface and the atmosphere, never coming close to a methanotroph.

Some methanotrophs require oxygen to degrade methane by aerobic oxidation. Only bacteria are known to carry out this reaction. During aerobic oxidation, methanotrophs harvest energy by transferring electrons from methane to oxygen, which transforms methane into carbon dioxide. Aerobic

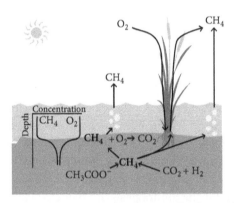

Figure 7.3 Release of methane from freshwater wetlands. Methane (CH_4) is produced by methanogens, here using carbon dioxide (CO_2) and hydrogen gas (H_2) or acetate (CH_3COO^-), and is released to the atmosphere by plants or via bubbles. The plant vascular system transporting oxygen (O_2) to roots provides a route for methane to travel up to the atmosphere. Methanogens are active only in the absence of oxygen, here in the sediments, while aerobic methanotrophs oxidize methane where both oxygen and methane are present, at the water-sediment interface.

methanotrophs must live at the interface where both oxygen and methane coexist. Oxygen diffuses down into this interfacial microhabitat from the atmosphere and from photosynthesizing microbes, whereas methane diffuses up from sediments where oxygen is absent and methanogens are active. Aerobic methanotrophs act as a biofilter that minimizes release of methane from underlying anoxic habitats into the atmosphere.

Another methane-degrading mechanism doesn't involve oxygen. It was discovered more than 50 years after oxygen-dependent methane degradation was documented and the first methanotrophic bacterium was isolated.[33] In the late 1970s, marine geochemists looked at how several chemicals varied with depth in coastal sediments and noticed that methane disappeared at the same place as sulfate did but in the absence of oxygen. The pattern suggested that sulfate was taking the place of oxygen in consuming methane, an indication of anaerobic oxidation of methane, unknown to geochemists at the time. If confirmed, it would be a new mechanism to degrade a powerful greenhouse gas in or near where it is produced in oxygen-free habitats.

Microbiologists also didn't know that methane could be oxidized anaerobically, so the prospect of discovering a new form of metabolism was exciting. A few years after the geochemists published their studies, microbial ecologists Alexander Zehnder and Thomas Brock, working then at the University of Wisconsin-Madison, found evidence of anaerobic methane oxidation in laboratory incubations that favored methanogenesis. They proposed that the oxidation was carried out by a symbiotic partnership between archaea, which use "reverse methanogenesis" to start the degradation process, and sulfate-reducing bacteria (Fig. 7.4). In 2000, Antje Boetius and colleagues at the Max Plank Institute for Marine Microbiology in Bremen, Germany, used molecular probes and sophisticated light microscopy to explore anaerobic oxidation of methane in California's Eel River Basin, the Hydrate Ridge off Oregon, and the Black Sea. They found cells of archaea and sulfate reducers bound together in tight aggregates. Other work demonstrated that the $\delta^{13}C$ of archaeal membranes was very negative, which can be explained only if the archaea degrade the gas and use methane carbon to make their membranes. Research over the last two decades has demonstrated that the symbiosis carrying out anaerobic oxidation consumes about 90 percent of the methane produced by methanogenesis in the oceans.

Methane is the unlikely foundation for other symbioses, including one first discovered in the Gulf of Mexico. While oil companies spend billions of dollars on hundreds of drilling rigs to extract natural gas from the Gulf,

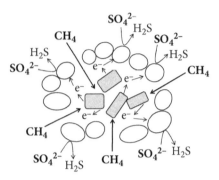

Figure 7.4 Anaerobic oxidation of methane. Archaea (shaded rectangles) carry out "reverse methanogenesis" and oxidize methane to carbon dioxide (not shown), releasing the electrons (e^-) to sulfate reducers (open ovals) which reduce sulfate (SO_4^{2-}) to hydrogen sulfide (H_2S). The cells are much closer together than depicted here.

methane naturally oozes out through many cracks, or cold seeps, in the Gulf's seafloor. In the early 1980s, the methane-oozing cold seeps attracted the attention of marine zoologists who were searching for more examples like the tubeworms discovered to be living near hot hydrothermal vents in the Pacific Ocean. Colleen Cavanaugh, then a graduate student at Harvard University, now a professor at Harvard, had discovered that the tubeworms depended on symbiotic bacteria using chemosynthesis to turn the reduced sulfur spewing out of vents into food for its invertebrate host. Rather than tubeworms, the marine zoologists searching the Gulf found dense communities of mussels and other invertebrates at methane cold seeps.[34] The mussels share many similarities with tubeworms but differ in one crucial aspect. Both are in oases of life surrounded by a barren ocean floor, far away from conventional food sources normally required to support large animal populations. The crucial difference was found to be their food, as revealed by ^{13}C analyses. Unlike tubeworms, mussels had very little ^{13}C (their $\delta^{13}C$ was very negative), which only made sense if their carbon came from methane.[35] But like all animals and other eukaryotes, mussels can't use methane. The authors of the ^{13}C study speculated that cold seep mussels fed on free-living methanotrophs that were growing on methane bubbling out from the seep. Fresh from her work on chemosynthetic symbiotic bacteria in tubeworms, Cavanaugh thought the methanotrophs may be symbionts living within the mussels. To support her hypothesis, she and colleagues used electron microscopy to document the

presence of bacteria within the mussel's gills and performed bioassays to uncover enzymes characteristic of aerobic methanotrophic bacteria.[36] Since
their discovery, mussels and their symbiotic methanotrophs have been found
at cold seeps around the world, including in the Arctic Ocean known for its
natural gas reserves. Similar symbiotic methanotrophs have been found in
Sphagnum moss submerged in peat bogs, another methane-rich habitat.[37] As
with free-living methanotrophs, the symbionts aren't 100 percent effective
in preventing the release of methane, but without them emissions would be
much higher.

Other methane-based symbioses include methanogens that make the gas
in partnership with mammals and insects. The mammals are cattle, sheep,
and other ruminants important in agriculture that, thanks to their symbiotic methanogens, produce more methane than the fossil fuel industry
does. Domesticated ruminants as well as wild ones, such as deer and ante-
lope, cannot digest the plant material they eat without help from a complex
community of symbiotic microbes, including methanogens, which decompose ingested plant material into fatty acids used for food by the animal.[38]
Methanogens are also in termites.[39] Like ruminants, termites depend on bacteria, protists, fungi, and methanogens (the exact composition depending on
the termite species) for turning wood ingested by termites into useable food.
The methanogen-termite symbiosis is important enough to warrant its own
subsection in the latest Global Methane Budget, although it is responsible for
only about 2 percent of all natural emissions.[40] Symbiotic methanogens have
also been found in cockroaches and scarab beetle larvae.

Methanogens and symbiotic bacteria are also in the organism responsible
for producing a lot of methane: *Homo sapiens*. Emissions directly from the
human body, however, are small compared to what comes out of cows, even
though as many as 60 percent of healthy people release methane through
belches and farts.[41] (Of course, humans have found too many other ways to
add methane to the atmosphere.) Rather than methane-spewing archaea,
our digestive tract is full of mostly bacteria, which along with other microbes
make up a collective known as the human microbiome. Our microbiome
is so huge that the human body is more microbial than human; an average
adult houses 38 trillion bacteria and only 30 trillion human cells.[42] The
human microbiome is an indispensable partner in maintaining our health
and well-being. In addition to the well-known ways by which microbes aid
in digestion, each day brings new discoveries of microbial ties with obesity,
neurological disorders, cancers, even our inclination to exercise.[43] Despite

the power of the human microbiome, however, it is not to blame for our contribution to climate change.

The human microbiome may not have much to do with methane, yet that is far from the case for methanogens in the rumen microbiome, which are a big reason why agriculture is such a large source of methane. Release of methane would be even higher if not for archaea related to methanogens that in partnership with sulfate-reducing bacteria degrade methane without oxygen. Symbioses and microbiomes would be studied by microbial ecologists even without climate change, but their work takes on an urgency because of the need to understand the rise in a powerful greenhouse gas, methane.

Chapter 9 has more about methanogens in cattle, manure piles, and rice paddies, but to complete this chapter and our survey of methane as a greenhouse gas, we need to return to natural environments where methane is produced by free-living methanogens. Recent studies about the recent rise in atmospheric methane have focused on the tropics and their wetlands and have not said much about the Arctic. In Chapter 3, we discussed the release of carbon dioxide from thawing permafrost in the Arctic. Another big concern is methane.

Burning Ice

According to Anton Chekhov's rule about loaded guns, this section shouldn't be here. Chekhov argued that if a loaded gun is introduced in a story, within a few pages it needs to go off. Otherwise, the gun is a distraction. In this chapter, the main story is about the atmospheric methane from agriculture and wetlands, all made by methanogens. Levels of this potent greenhouse gas are increasing, not only because of fossil fuels, but also because we are raising more livestock and cultivating more rice and because changes in the hydrological cycle have stimulated more methane release from wetlands. Methane hydrates are the loaded guns unloved by Chekhov. They aren't going off today, as hydrates currently aren't big sources of atmospheric methane, yet they are more than just a distraction. Although the exact number isn't known, even the lowest estimate of how much methane is tied up in hydrates is incredible. Hydrates may contain 2,000,000 to 8,000,000 teragrams of methane, nearly 500 to over 1,500 times the amount in the atmosphere, according to the latest IPCC report. Hydrates may house as much as five times the amount of methane in known natural gas reserves, which hasn't escaped the attention of petroleum companies. (Because of technical problems, however, commercial production of natural gas from hydrates is still in its infancy.)

If Franklin and Washington were amazed to see inflammable air bubbling up from New Jersey mud, they would have been even more astounded to learn that a form of ice can be set on fire. A chunk of methane hydrate, also called methane clathrate, looks like ice because that's close to what it is: a few methane molecules surrounded by several frozen water molecules. The most common form of hydrates has eight methane molecules encased in a frozen cage of 46 water molecules.[44] (Other gases can be the "guest" within a hydrate's water-ice cages.) Despite the seemingly low amount of methane in each cage, a liter of hydrate holds the equivalent of nearly 200 liters of methane at room temperature and sea level pressure.[45] When a piece of melting hydrate is ignited, blue flames eerily rise out of white ice.

Methane hydrates form and stay frozen only under special conditions. The temperature must be low, or the pressure high, or a combination of both. Although atmospheric pressure is very low at the surface of Mars, it's cold enough to have methane hydrates.[46] The natural-gas extraction industry has a big problem when pressures reach a high enough level in underwater pipes for hydrates to form. (Methane hydrates were first discovered in the 1930s in a pipeline.[47]) There has been speculation that the Deepwater Horizon blowout on April 10, 2010, which killed eleven men and devastated the Gulf of Mexico ecosystem, was caused by a methane hydrate.[48] When it comes to climate change, the concern is the melting of hydrates. The worry is even a small change in temperature could melt them and release methane that would exacerbate global warming. The power of hydrates and methane can be seen in the geological past.

Methane hydrates have been cast as either the hero or the villain in stories about past climate change. Hydrates may have pulled Earth out of ice ages so severe that glaciers completely smothered the planet even in the tropics. One of these extreme glacial periods, known as Snowball Earth, occurred in the Neoproterozoic about 635 million years ago.[49] One hypothesis is that the melting of methane hydrates caused enough global warming to end the Neoproterozoic Snowball Earth and to create a warmer, ice-free climate more amiable to life; biological diversity exploded as a result. A piece of evidence implicating methane is the low ^{13}C content of carbonates deposited at the time, made with ^{13}C-depleted carbon dioxide left over from the oxidation of methane released from melting hydrates. (Here and in the other cases discussed next, most of the methane released from melting hydrates probably was degraded to carbon dioxide before reaching the atmosphere, but the end result was the same: global warming.) Rather than rescuing Earth from being smothered in ice, ^{13}C data point to methane as causing too much global warming at other times in the geological past. That may have been the case at the end of the early Jurassic about 183 million years ago and during the Paleocene-Eocene Thermal Maximum, about 55 million years ago when the planet warmed by 5 to 8 degrees Celsius. As with the other climate-change episodes, methane released from melting hydrates would explain the warming and the low ^{13}C values, although other mechanisms, such as high volcanic activity, have been proposed. Warmer, oxygen-poor deep waters caused the extinction of many benthic organisms while the diversity of terrestrial mammals expanded.[50] Some of the best evidence implicating methane hydrates in ending an ice age comes from the Oligocene–Miocene boundary 23 million years ago.[51] Getting closer to our time, in the Late Pleistocene, about 130,000 to 12,000 years ago, the Clathrate Gun hypothesis may explain the short periods of warm temperatures that interrupted longer periods when glaciers covered northern Europe and North America. According to the hypothesis, slight warming of oceanic waters, perhaps due to changes in the ocean conveyor belt, melted hydrates along continental slopes and released enough methane to warm the planet briefly before the ice age returned.[52] With all stories from the geological past, there are other explanations for the temperature swings, the ^{13}C data, bursts in biological diversity, and the mass extinctions. Yet, even if not the complete explanation, methane hydrates are likely to be a big reason why Earth went through several large climate changes over eons.

The question is, what about now and the future? To understand more clearly the potential for hydrates to effect climate change today, we need to look at where they occur and their susceptibility to melting by global warming.

The biggest region with the right conditions for methane hydrates to exist is at the bottom of the ocean.[53] The right conditions are a combination of pressure, thus water depth, and temperature. Hydrates are only possible in temperate latitudes if the water is at least 500 meters deep, whereas in the Arctic Ocean where bottom waters are colder the limit is about 300 meters. (Not enough is known about methane hydrates in Antarctica to say more here.) Hydrates in the deepest oceans, greater than 1000 meters, are highly unlikely to release much methane to the atmosphere over the next 100 years. Heat from global warming needs a very long time, millennia, to penetrate the deepest sediments, and if deep waters heat up more than expected, sea level would also rise and the water pressure, which keeps methane hydrates solid, would increase. A final barrier stopping methane from reaching the atmosphere is methanotrophy. Methane hydrates buried beneath the ocean floor are under a layer of sulfate-rich sediment several meters thick where anaerobic methanotrophy is active. If a molecule of methane somehow escapes the anaerobic methanotrophs, it will not likely survive the trip through hundreds of meters of water filled with aerobic methane degraders. We should be more worried about other problems.

One potential problem is hydrates in coastal sediments in the Arctic. The methane in these hydrates mostly comes from methanogenesis fueled by organic matter sinking from productive phytoplankton at the surface and from terrestrial runoff. Being at the edge of the temperature-pressure boundary of hydrate stability, shallow methane hydrates in the Arctic can be melted by a warm oceanic current passing by or even by small changes in pressure associated with tides. Global warming was initially thought to have triggered the release of methane from melting hydrates in coastal sediments off Svalbard,[54] the Norwegian archipelago north of the Arctic Circle mentioned in Chapter 6. Subsequent work by Christian Berndt and colleagues at the Helmholtz Centre for Ocean Research in Kiel, Germany, established that methane had been leaking from Svalbard sediments for at least 3000 years and that bottom waters above the hydrates normally fluctuate a degree or two. Warmer bottom waters would melt the hydrates, but when temperatures dropped, the hydrates grew back. Berndt and colleagues argued the recent release of methane from Svalbard sediments wasn't due to recent warming. But they acknowledged that temperatures are trending up in coastal Svalbard waters as global warming continues and that periods of hydrate melting will last longer and longer. So it sounds like future global warming could melt hydrates in the coastal Arctic Ocean and release lots of methane.

Fortunately, that's unlikely to happen, at least for the next several decades. Several modeling studies have found that little of the huge methane hydrate reservoir frozen in the Arctic Ocean is likely to melt over the next hundred years, and any melting will not add much methane to the atmosphere.[55] One study calculated that melting hydrates in the Arctic would release only 4.73 teragrams of methane each year, which is very small compared to total anthropogenic emissions of about 360 teragrams per year. Granted, the 4.73 teragrams estimate is probably too low because the study used a low number for how much methane is contained in hydrates. But the study didn't consider degradation by aerobic and anaerobic methanotrophs, which can be very effective in stopping methane from reaching the atmosphere. It is reassuring to note that even though methane made up nearly a third by weight of the petroleum-like chemicals leaked during the Deepwater Horizon blowout, most of it was degraded by aerobic methanotrophs in the water column before it could reach the atmosphere.[56]

There is one more place in the Arctic with methane hydrates: on land buried deep beneath permafrost. Even if these hydrates are like the others, another Chekhov gun that isn't likely to go off, we still should consider the possibility that they may melt in the future. In Chapter 3, we saw that permafrost above these hydrates is already thawing and releasing carbon dioxide from the degradation of thawed organic material by microbes. Another concern is thawing permafrost could release a more powerful greenhouse gas, methane.

Once and Future Methane in the Arctic

Michael Adams, a young biologist at the Russian Academy of Sciences, traveled to the shores of the Bykovskiy Peninsula in Siberia when he learned what Tungusic chief Shoumakhov had found in 1799 in a chunk of ice that had separated from a cliff. As the ice melted, Shoumakhov saw two feet of what we now know was a mammoth.[57] By the time Adams got to the carcass, the flesh and internal organs had been eaten by wild animals, but he was able to salvage the skeleton and bring it to Saint Petersburg where it is now on display at the Museum of Zoology. Adams's publication describing the mammoth was a worldwide sensation, even though it wasn't the first. He was one of the first, however, to describe yedoma, the combination of frozen soil and ice that preserved the mammoth for several thousands of years. Formed during the Pleistocene, yedoma is a type of organic, ice-rich permafrost up to 40 meters thick that covers about 1 million square kilometers in present-day Siberia and Alaska.[58] Although yedoma occurs greater than 3 meters below the tundra surface, other types of permafrost can be much closer to the surface. All types are overlaid by an active layer that warms up enough in summer for microbes to function and grow. Below or sometimes within the permafrost layer, methane hydrates are possible (Fig. 7.5).

According to the latest IPCC report, permafrost contains about 500,000 teragrams of methane. That's tiny compared to the methane hydrates in ocean

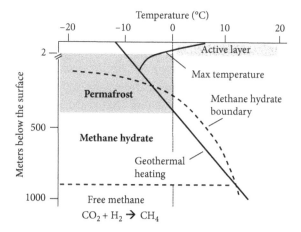

Figure 7.5 Permafrost and methane hydrates in the Arctic. The depths for the active layer (where temperatures are warm enough for microbes to be active), permafrost, and methane hydrate vary among locations. Methane hydrates are potentially found below the curved dashed line ("Methane hydrate boundary"), as shallow as 200 meters for temperatures about −10 degrees Celsius, overlapping with permafrost. Hydrates can extend down to the horizontal dashed line, about 800 meters, even though geothermal heating warms temperatures above freezing. Below hydrates, free methane can be produced by methanogenesis, or it can come from geological sources. Based on Ruppel and Kessler (2016).

sediments, but it's over 10 times the amount of methane now in the atmosphere. Along with those on land, permafrost and associated hydrates can be found under shallow coastal waters of the Arctic Ocean. (Hydrates are also under Lake Baikal and on the Tibetan Plateau, but they're too small to worry about.) The submerged permafrost in the coastal Arctic Ocean originally formed on land tens of thousands of years ago but then was inundated by water when the sea level rose as the last ice age ended. Below the permafrost layer, methane continues to be produced where temperatures are warm enough for methanogens and other microbes to be active, fueled by ancient organic chemicals left over after being buried by soil that had accumulated over millennia. (Geothermal heating causes temperatures to rise with depth below the land surface.) Additional methane comes from deeper, geological sources. Free methane from both sources can't escape to the atmosphere as long as permafrost and hydrates remain as frozen caps. The burning question is how much of the caps is thawing today and how much more will thaw in the future?

We know large sections of permafrost are already thawing as the Arctic warms, nearly four times faster than the rest of the planet has.[59] Some consequences are obvious, as mentioned in Chapter 3. Buildings slump, roads buckle, and coastal cliffs along the Arctic Ocean collapse when their permafrost base thaws. Thawing permafrost has also released large amounts of carbon dioxide, but what's happening with methane is less clear. Just as in the deep ocean, global warming must penetrate many meters of soil and rock before reaching methane hydrates below the permafrost layer, and any methane released from a melting hydrate then would have to survive the journey through the methanotroph filter before reaching the atmosphere. Even though the environs around Utqiaġvik, a small village perched on the northern coast of Alaska, are warming like everywhere else in the Arctic, 30 years of monitoring hasn't detected any increase in methane emissions,[60] and the latest Global Methane Budget doesn't include any methane emitted from hydrates of all sorts, including those associated with permafrost. An exception may be permafrost hydrates in Siberia, but the high emission rates initially reported for the region have not been confirmed by subsequent studies.[61] In short, the available data indicate permafrost hydrates are not a large source of atmospheric methane today and seem unlikely to release much methane over the next hundred years.

But it's too early to cross off permafrost from our list of things to worry about. A problem in getting a definitive answer about permafrost and their hydrates is the many difficulties in studying them in such a large, complex, and harsh ecosystem in remote corners of the planet. Permafrost coverage is complicated and uneven, and the extent of underlying hydrates is uncertain. With global warming, small, hard-to-detect hotspots of methane emerge.[62] Further complicating the picture are sudden changes in how fast permafrost thaws. Large-scale models of methane emissions now simulate gradual changes, but not sudden ones, such as the erosion of hills and cliffs and the appearance of slumps, gullies, and small ponds and lakes where water accumulates in depressions, known as "thermokarsts," formed when permafrost quickly thaws.[63] As the Arctic landscape is transformed by rapidly thawing permafrost, so too is the carbon cycle. A modeling study led by Merritt Turetsky suggests that even though abrupt thawing may hit less than 5 percent of permafrost, it accounts for about 40 percent of the carbon emissions attributed to gradual thawing. It would be simpler if the emissions co-occurred with the thawing, but organic carbon released by abrupt thawing can be carried downslope by erosion and buried again or oxidized elsewhere by microbes. The Turetsky study points out that as gradual thawing works its way through soil centimeter by centimeter over decades, abrupt thawing may transform meters of permafrost in days.

The speed at which permafrost thaws may affect what microbes do with the released organic material. Along with freeing methane trapped in or below permafrost, abrupt thawing may change the relative amounts of methane versus carbon dioxide released during organic material degradation. Most studies conclude that carbon dioxide is the main greenhouse gas released when organic material frozen in permafrost thaws and becomes available to oxidation by bacteria and fungi. A meta-analysis of 25 incubation studies found that when permafrost was experimentally warmed, release of carbon dioxide was much higher than that of methane, so much higher that even after accounting for the fact that methane is a 20 times more powerful greenhouse gas, carbon dioxide release had a larger impact.[64] But methane may have more of an impact with abrupt thawing. In the Turetsky model, during abrupt warming, methane accounted for about 20 percent of carbon emissions but about half of the total greenhouse effect of all gases released during degradation. When permafrost thaws quickly, soil becomes waterlogged and loses oxygen as aerobic microbes degrade the thawed organic material. The now anoxic soil is ripe for methanogenesis. Sediments beneath thermokarst lakes are also anoxic and methane hotspots. Thawing permafrost helps to explain why 24 percent of the recent increase in atmospheric methane was due to emissions from Canada and Alaska.[65] Abrupt thawing is another climate-change wild card. As the Arctic continues to warm, methane emissions from permafrost and methane hydrates may become bigger and unfortunately easier to quantify. Much remains unknown at this time.

What is clear is the central importance of microbes in governing levels of a powerful greenhouse gas. Although fossil fuels are a large source, even more methane comes from methanogens in wetlands, ruminants, and rice fields. Still, humans are responsible for the rise in atmospheric methane, including the methane coming from wetlands. From swamps in New Jersey to marshes in Italy, wetlands have released a lot of methane over the last decade because of global warming caused by human activity. While some microbes contribute to the problem, others work at mitigating it. The atmosphere would have even more methane if not for degradation by bacteria with oxygen and by the archaea-bacteria symbiosis without oxygen. What microbes do is essential for understanding methane and its contribution to climate change.

8

Laughing Gas

Nitrous oxide is more than just another greenhouse gas. It's used as an analgesic in dentistry and as an antidepressant in psychotherapy,[1] and it appeared as a prop in the first installment of the action-movie franchise *The Fast and Furious*. For the movie's first street race, both protagonists, an undercover cop, Brian O'Conner, and the leader of a car hijacker gang, Dom Toretto, have tricked out their cars with tanks containing nitrous oxide.[2] The gas gets more power out of car engines in real life as well as in Hollywood. As it breaks down at high temperature, nitrous oxide in effect delivers concentrated oxygen gas for engines to burn fuel more efficiently. The same chemistry explains nitrous oxide's utility as a rocket propellant. Meanwhile in homes and restaurants, nitrous oxide takes the place of a beater in the making of whipped cream. The gas is also a recreational drug that gives the partaker a quick buzz when inhaled, which is the reason why the state of New York outlawed the sale of nitrous oxide whipped cream chargers, known as whippets, to people under 21.[3]

Nitrous oxide was used as a drug, for both medicinal and recreational reasons, long before it appeared in a Hollywood movie. The gas was discovered in 1772 by Joseph Priestley, that English chemist who corresponded with Benjamin Franklin about inflammable air (Chapter 7), but one of Priestley's compatriots and colleagues, Humphry Davy, gave it the name used today (Priestley had called it dephlogisticated nitrous air) and discovered its mind-blowing properties.[4] Along with writing poetry, ballads, metaphysical essays, and a book about fly fishing for salmon, Davy had the intellect, energy, and ambition to make his mark on the nascent field of chemistry, which he and others thought would reveal the true nature of the material world.[5] He also worked on practical problems, such as the potential healing power of gases. Before exposing his patients to new treatments, he tried them out on himself, often with severe consequences. His experiment with carbon monoxide, a toxic gas emitted by cars and portable generators, nearly killed him. Undeterred, he moved on to nitrous oxide. After figuring out which chemical reactions were best to produce the gas, Davy tested nitrous oxide out on a small dog, a "healthy kitten of about five months old," and himself.[6] On December 26, 1798, Davy sat inside an airtight box with 20 quarts of nitrous oxide. He felt nothing for three minutes, and then his cheeks reddened, his pulse increased, and his temperature rose. After 30 minutes, 20 more quarts were added to the box, and then later, another 20. At this point he felt a "great disposition to laugh,"[7] and after recovering he exclaimed that he felt "nothing exists but thoughts!—the universe is composed of impressions, idea, and pleasures and pain."[8] He introduced nitrous oxide to several influential friends in London, and soon gatherings imbibing the gas were called "laughing gas" parties. Davy recognized the analgesic effect of nitrous oxide, but it wasn't tested on a patient until decades later, in 1844, and it didn't become widely used until the 1860s.[9] Although too weak to be suitable for major surgeries, it is used today along with a local anesthetic in dentistry.

Unfortunately, nitrous oxide doesn't stay in the dentist's office, much less in car engines, rockets, or whippets. As mentioned in Chapter 1, it's the third most important greenhouse gas in the atmosphere, after carbon dioxide and methane, being much less common than the other two greenhouse gases, with levels about 1000 times lower than that of carbon dioxide. Despite its low atmospheric levels, nitrous oxide is troubling because each molecule is about 300 times and over 10 times more powerful than molecules of carbon dioxide and methane are, respectively, in trapping heat over a 100-year time horizon.[10] Data about atmospheric levels come from the same sources we've already seen for carbon dioxide and methane, including a global network of monitoring stations that started in 1995, which was later than for the other gases.[11] Data from ice cores at Law Dome, Antarctica, show nitrous oxide levels going down and up with temperature as ice ages came and went over the last 800,000 years.[12] After varying around 270 parts per billion for several millennia, levels began to increase in the late nineteenth century, and the rise has been especially big over the last five decades. Unlike methane which paused before continuing its upward surge (Chapter 7), atmospheric nitrous oxide has been increasing relentlessly. Unfortunately, nitrous oxide is more than just a potent greenhouse gas. It also destroys more ozone than chlorofluorocarbons (CFCs) now do, thanks to the success of the Montreal Protocol in regulating CFCs.[13] Later we'll see that nitrous oxide's rise in the atmosphere is a harbinger of other environmental problems.

The recent increase in atmospheric nitrous oxide should sound familiar, as it's like what we've seen for carbon dioxide and methane. Levels of all three greenhouse gases bounced up and down for hundreds of thousands of years, and then in the nineteenth century, they began to surge to unprecedented levels.[14] It should come as no surprise that nitrous oxide levels, like those of the other two greenhouse gases, have risen because of human activity. The immediate cause for the rise is easy to see. It will take a bit of work to uncover the deeper, underlying reason. Nitrous oxide's story differs in many ways from those of carbon dioxide and methane, starting with its chemical constituents (N_2O); with no carbon atoms, it's in the nitrogen cycle, which is run by different types of microbes and microbial processes than we've encountered so far.

The Hidden Source

A starting point for understanding the increase in atmospheric nitrous oxide is to look at its sources. In Chapter 7, we saw that natural sources are big emitters of methane, roughly equal to the amount released by agriculture and the burning of fossil fuels. Likewise for nitrous oxide, natural sources release a bit more than anthropogenic ones do (Fig 8.1), but unlike methane, the ocean is a large source, emitting only slightly less nitrous oxide than what comes out of grassland and forest soils.[15] Despite being big sources, however, there is no indication that soils and the oceans are responsible for the recent rise in this powerful greenhouse gas.

Surprisingly, the rise isn't due to fossil fuels, at least directly. Each year the burning of oil, coal, and natural gas releases only about one teragram of nitrogen (Tg N), a small fraction of the over 8 Tg N from all anthropogenic sources, tiny compared to the nearly 10 Tg N from natural sources. As for methane, agriculture again is the biggest source by far, emitting much more nitrous oxide than fossil fuel combustion does. Another large anthropogenic source is labeled "atmospheric deposition" in Figure 8.1, consisting of nitrogen chemicals carried by precipitation ("wet deposition") or by big enough particles that fall out of the atmosphere ("dry deposition") onto land and aquatic habitats. The nitrogen chemicals include ammonia (NH_3) emitted by agriculture and nitrogen oxides (mainly NO and NO_2) released by vehicles, power plants, and factories. One recent study using nitrogen isotope data concluded that atmospheric deposition is as much as half of all anthropogenic sources.[16] So fossil fuels may be responsible for a bit more of the nitrous oxide than the 1 Tg N per year indicated in Figure 8.1, but the estimate for agriculture should also be bumped up. In any event, atmospheric

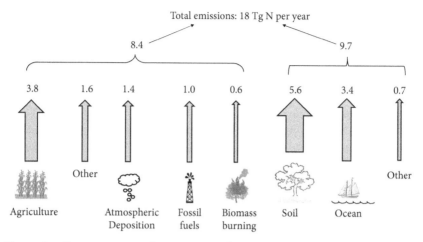

Figure 8.1 Emission rates of nitrous oxide from anthropogenic and natural sources in teragrams of nitrogen (Tg N) per year. "Biomass burning" refers to the burning of forests and grasslands caused by human activities. For "Atmospheric deposition," nitrogenous chemicals are carried by precipitation or on particles big enough to fall out of the atmosphere onto soils, freshwaters, or the oceans where the chemicals are transformed to nitrous oxide. Data from H. Tian and colleagues (2020).

deposition doesn't change the conclusion that agriculture releases more nitrous oxide than any other human activity.

Agriculture accounts for nearly 90 percent of the rise in atmospheric nitrous oxide over the last four decades.[17] As discussed in Chapter 7, agriculture is also a big source of methane, but where and how agriculture releases the two gases differ in many respects. Some nitrous oxide is emitted along with methane in belches and farts from cattle and other ruminants, but that's minor compared to other sources. More is released from manure, accounting for nearly 40 percent of all nitrous oxide coming from agriculture. The biggest anthropogenic source, the other 60 percent attributed to agriculture, is cropland soil. As will be discussed in more detail later, soil microbes transform a variety of nitrogenous inorganic and organic chemicals to nitrous oxide, which then escapes into the atmosphere. The chemicals, the microbes, and the transformations occur in all soils, those in natural habitats as well as on farms. The difference is farm soils are fertilized.

Nitrogen fertilizer is the main reason for the rise in atmospheric nitrous oxide, especially so over the last six decades. More and more fertilizer has been used over the years, made possible by a change in the kind of fertilizer available to farmers as agriculture evolved from small farms to a large agribusiness. Since the beginning of agriculture in the Neolithic Age, farmers have used organic nitrogen, aka manure, to fertilize cereal crops and pulses.[18] In the mid-nineteenth century, manure, the remains from meatpacking plants, and other forms of organic nitrogen continued to be used in commercial fertilizers. Guano was an important organic fertilizer for a few decades. Although bird droppings, the main stuff of guano, are everywhere, only in a few places in the world are conditions right for the buildup of guano deposits big enough to be commercially exploited. During the Guano Age from 1840 through 1880, guano dominated the commercial fertilizer market,[19] reaching a peak of 600,000 tons in the late 1860s, but by start of the twentieth century as the deposits were depleted, guano made up only a few percent of the fertilizer market.[20] Another commercial nitrogen fertilizer, sodium nitrate, went through a similar boom and bust about the same time as the Guano Age. Like guano, deposits of sodium nitrate, the biggest being in Chile, were not large enough to satisfy the worldwide demand for long.

As the nineteenth century came to a close, scientists and leaders of the Great Powers were worried that the world was running out of nitrogen. In his 1898 presentation to the British Association for the Advancement of Science, Sir William Crookes argued "England and all civilized nations stand in deadly peril of not having enough to eat,"[21] because of the looming nitrogen fertilizer shortage. The Great Power leaders were also worried about running out of the nitrogen needed to make munitions; potassium nitrate is a major ingredient of gunpowder. In 1900, about half of the sodium nitrate imported by the United States was used to manufacture explosives. Germany, the largest importer at the time, was especially concerned about access to sodium nitrate because the country's oversea supply could be easily choked off by a naval blockade. The problem was attacked by world-leading chemists working for BASF and other powerful chemical companies in Germany.

The problem was solved by two Germans, Fritz Haber and Carl Bosch, who figured out how to make nitrate from the biggest source of nitrogen, nitrogen gas, which makes up about 78 percent of air. In 1909, Haber discovered how to synthesize ammonia from nitrogen and hydrogen gases, and in a few years Bosch scaled up Haber's laboratory experiments to massive ammonia-generating factories. With ammonia, chemical engineers could make sodium nitrate and all other nitrogenous chemicals needed for agriculture as well as for industry and the military. The Haber-Bosch process replicates what microbes do at room temperature (or lower) and normal atmospheric pressure but at a much higher temperature (over 400 degrees Celsius) and pressure (over 140 atmospheres).[22] The Haber-Bosch process needs the extreme conditions to break the triple bond holding the two atoms together in nitrogen gas (N_2) and to combine a nitrogen atom with hydrogen gas (H_2). The triple bond of nitrogen gas is stronger than a single bond such as in hydrogen gas or even the double bond found in oxygen gas and carbon dioxide. The process of converting nitrogen gas to ammonium is called nitrogen fixation because it transforms, or fixes, nitrogen gas into a nongaseous chemical. In nature, only a few bacteria and archaea are able to carry out nitrogen fixation. It's an energy-intensive process, requiring a very large, complex enzyme, nitrogenase. A few cyanobacteria are successful nitrogen fixers in part because they can harness light energy. No eukaryote can fix nitrogen. The nitrogen fixation carried out by legumes such as soybeans is actually done by their symbiotic nitrogen-fixing bacteria.

While microbes use the energy they get from light or organic material oxidation to fix nitrogen gas, the Haber-Bosch process currently relies on the energy in fossil fuels. Nitrogen fertilizer production consumes about 5 percent of the global natural gas supply, while coal is the main energy source in China.[23] Along with the methane used to make the hydrogen gas needed in the Haber reaction, fossil fuels are also needed to achieve the high temperature and pressure required by the Haber-Bosch process. Without fossil fuels, it's unlikely that the huge increase in nitrogen fertilizer production during the twentieth century would have been possible. Perhaps other sources of energy such as hydroelectricity could have been harnessed, but they wouldn't have been as cheap in the 1900s or as pervasive as fossil fuels were. So the inexpensive, abundant, and widely available energy provided by fossil fuels was a key to the rise in fertilizer nitrogen production and eventually atmospheric nitrous oxide over the last 75 years. Agriculture may be the biggest anthropogenic source of nitrous oxide, but in many ways, the real biggest source is fossil fuels.

Too Much of a Good Thing

The Haber-Bosch process has been for good for humanity (ignoring the nitrogen that has gone into munitions) but bad for the environment. Thanks in part to the nitrogen fixed by the Haber-Bosch process, agricultural productivity has been increasing since the mid-twentieth century, helping the world support a burgeoning human population if only imperfectly. Even with the Haber-Bosch process, too many people face food insecurity and go hungry, but it could be worse. Vaclav Smil has argued that there would be three billion fewer people on the planet without Haber-Bosch.[24] Because of the pressing need for nitrogen fertilizer, the amount of nitrogen fixed by the Haber-Bosch process is unbelievably large, roughly equal to the amount fixed in nature by microbes; the input of nitrogen into the biosphere has doubled over the last 100 years. By comparison, the carbon released by fossil fuel burning is less than 10 percent of natural inputs (Chapter 2), yet even that relatively low percent is having a huge impact on Earth's climate. The impact of so much nitrogen coming from nitrogen fertilizer is also becoming clear.

A sign of those impacts is the rise in atmospheric nitrous oxide over the last 75 years. Nitrous oxide rose only slowly during the first few decades following the invention of the Haber-Bosch process when use of nitrogen fertilizer remained flat, as agricultural productivity was held back by the Great Depression and other economic headwinds. Following World War II, however, nitrogen fertilizer application took off as did atmospheric nitrous oxide (Fig. 8.2). Although fertilizer use in the United States and western Europe leveled off in the 1960s, it has increased substantially in China and the rest

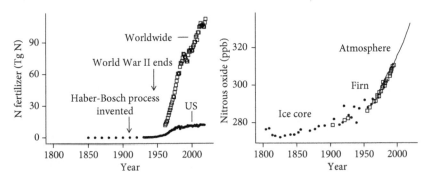

Figure 8.2 Nitrogen fertilizer application and atmospheric nitrous oxide over time. The fertilizer application data for the United States and worldwide are from the US Department of Agriculture and the International Fertilizer Industry Association, respectively. Estimates of atmospheric nitrous oxide based on bubbles trapped in ice cores and firn (compacted snow) in Antarctica are from M. Battle and colleagues (1996), and the direct atmospheric measurements are from X. Lan, K.W. Thoning, and E.J. Dlugokencky at the NOAA Global Monitoring Laboratory.

of eastern Asia, so global use has continued to rise.[25] The problem is, plants don't take up all fertilizer spread on fields. The relative amount of applied fertilizer taken up by plants, or the nitrogen use efficiency, can be estimated by comparing the nitrogen in harvested crop plants with the amount of fertilizer nitrogen applied to the crop field. The nitrogen use efficiency varies greatly among regions and crop plants. The average efficiency for all crops using nitrogen fertilizer varies from as low as 25 percent in China (subsidies have encouraged excessive fertilizer application) to as high as 68 percent in the United States and Canada, while the worldwide average for specific crops can be as low as 14 percent for fruits and vegetables to as high as 46 percent for corn (maize).[26] The unused nitrogen fertilizer can stay in the field for a long time if it is incorporated into soil organic matter, or it can be transformed by microbes into nitrogen gas or nitrous oxide and escape into the atmosphere. Another fate is to be carried away by runoff into nearby creeks, streams, and rivers, eventually flowing into lakes and coastal oceans. Once in those waters, the fertilizer can cause several environmental problems. Along with contributing to the greenhouse effect, the rise in nitrous oxide is a symptom of nitrogen pollution.

Fertilizer nitrogen in aquatic habitats is a case of too much of a good thing. By the time it reaches a river or lake, fertilizer nitrogen has been converted to nitrate, which is a major nitrogen source for algae and cyanobacteria, the main starting point for food chains in all aquatic habitats, from small ponds to the oceans. Because nitrogen is often the element driving the growth of these plant-like microbes, more nitrate usually means more algal or cyanobacterial growth and biomass. With more growth at the bottom of the food chain comes more growth at the top in the form of fish and shellfish, as well as birds and terrestrial animals that depend on aquatic life for subsistence. That's all good and natural.

But too much nitrate turns it into a pollutant. High nitrate levels in drinking water may cause blue-baby syndrome (methemoglobinemia) and increase the risk of bladder and ovarian cancers,[27] while in aquatic habitats high levels can promote excessive algal and cyanobacterial growth, which color surface waters a sickly green. If the alga or cyanobacterium is toxic, the excessive growth is called a harmful algal bloom.[28] Toxins produced by algae such as *Heterosigma akashiwo* and *Aureococcus anophagefferens* can kill fish and other aquatic organisms and sicken even animals on land. Consumption of shellfish feeding on harmful algae can cause neurotoxic shellfish poisoning, amnesic shellfish poisoning, and diarrhetic shellfish poisoning.

When the cyanobacterium *Microcystis* bloomed in Lake Erie in 2014, the city of Toledo, Ohio, was forced to use other sources of drinking water to avoid the cyanobacterium's toxin, microcystin, which causes abdominal pain, diarrhea, and fever.[29] Some forms of microcystin are even carcinogenic.

Even if they aren't toxic, overly abundant algae and cyanobacteria create problems when they sink to bottom waters and are decomposed by bacteria. Depending on the amount of sinking organic matter from algae and cyanobacteria and how fast bottom waters are mixed with well-oxygenated surface waters, decomposition by bacteria can deplete dissolved oxygen to dangerously low levels.[30] Without oxygen, benthic animals eventually die, and fish trapped in low-oxygen waters wash ashore dead. Aquatic ecologists call these oxygen-poor waters hypoxic, and if oxygen is completely gone, anoxic. The popular press calls them "dead zones," a term that captures the potential destruction caused by low oxygen. Although hypoxic and anoxic waters have always naturally occurred, dead zones have multiplied worldwide and have become more severe over the last 75 years. The complete story for this expansion is complicated and involves other factors like phosphorus nutrients, wastewater treatment plants, and the physics of aqueous environments, but nitrogen fertilizers deserve much of the blame. The case against fertilizer is especially clear for the dead zone in bottom waters of the northern Gulf of Mexico. Work by Nancy Rabalais, Gene Turner, and their colleagues have demonstrated that increasing use of fertilizer in the U.S. Corn Belt in the twentieth century led to high nitrate levels in the Mississippi River, which eventually drains into the northern Gulf. Nutrients brought by the Mississippi have produced a dead zone the size of New Jersey, the second biggest worldwide. Fertilizer is the biggest source of nutrients fueling algal blooms and dead zones not only in the Gulf but also in the Chesapeake Bay, the Baltic Sea (the biggest dead zone), the northern coastal waters of the Black Sea, and elsewhere around the world.

Nitrous oxide doesn't cause harmful algal blooms and dead zones, yet they are connected, and there is a good reason to emphasize that connection. The general public can neither see nor probably appreciate the problems caused by atmospheric nitrous oxide, but everyone can see when a local reservoir, lake, or coastal bay is polluted. The increases in nitrous oxide, harmful algal blooms, and dead zones all can be traced back to fertilizer overuse. If atmospheric nitrous oxide levels continue to rise, it's a sign we will continue to have problems caused by nitrogen pollution. These problems worsen the environmental health of aquatic habitats and detract from the quality of life on land, including where we live, work, and play.

How to Make Nitrous Oxide

We need to get into some of the details of how microbes produce nitrous oxide if we are to understand the future of this potent greenhouse gas. Our tour of nitrous oxide production is also an opportunity to learn more about the ingenious ways bacteria and archaea make a living. Along with photosynthesis and respiration, the only forms of energy-generating metabolism carried out by large, supposedly more sophisticated organisms, microbes also take advantage of just about every energy-generating reaction that is permitted by the laws of thermodynamics. In Chapter 7, we saw how some archaea gain energy to support their growth by producing methane while other bacteria and archaea get the energy they need by decomposing methane. Chapter 6 mentioned that some bacteria in microbial mats use sulfate in place of oxygen, and other bacteria at hydrothermal vents live on the sulfur coming out of hydrothermal vents. All these alien ways of generating energy are known only to bacteria and archaea.

Nitrous oxide is at the center of several other alien pathways. The main ones that produce the gas are denitrification and nitrification, two key parts of the nitrogen cycle that starts with the fixation of nitrogen gas into ammonia (Fig. 8.3). Ammonia (NH_3) is quickly converted to ammonium (NH_4^+) at near-neutral pH and is taken up by plants and microbes to make protein, DNA, and other cellular components. After ingesting the plant material or microbes, herbivores and microbial predators excrete feces, urine, or urea—organic nitrogen which bacteria and fungi turn back into ammonium. Some of that ammonium can be converted into nitrate during nitrification, and that nitrate is then transformed into nitrogen gas during denitrification, completing the cycle. When operating optimally, denitrification and nitrification don't produce much nitrous oxide, so the gas often isn't included in illustrations of the nitrogen cycle. Unfortunately, the two processes don't work optimally all the time and leak out nitrous oxide.

Of the two reactions producing nitrous oxide, denitrification was discovered first, about the same time in the late nineteenth century as scientists and politicians began to worry about running out of nitrogen for fertilizer and munitions.[31] An obscure reaction known only to microbiologists suddenly seemed to threaten agriculture with troubling implications for society.[32] Denitrification was seen as a threat at the time, and remains a problem in agriculture today, because it converts nitrate to nitrogen gas or nitrous oxide, releasing valuable fertilizer nitrogen into thin air. When oxygen runs out and reaches low enough levels, nitrate takes its place so that denitrifying microbes

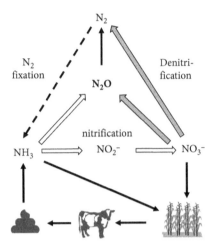

Figure 8.3 The nitrogen cycle. One type of nitrification (the white arrows) converts ammonia (NH_3) to nitrite (NO_2^-) and then to nitrate (NO_3^-). Denitrification (the gray arrow) transforms nitrate to nitrogen gas (N_2). Both nitrification and denitrification can produce nitrous oxide (N_2O). Two pathways mentioned in the chapter not depicted here are nitrification carried out by a single microbe ("comammox") and anaerobic oxidation of ammonium to nitrogen gas.

can degrade or, more precisely, oxidize organic chemicals and grow; the microbes gain energy when the electrons released or donated during organic matter oxidation go to nitrate, the electron acceptor if oxygen isn't present. Microbes prefer oxygen over nitrate as an electron acceptor because they can gain more energy using it, explaining why adding oxygen stops denitrification. It also explains why oxygen is so important in thinking about nitrous oxide production and degradation, as we'll soon see. Because the energy gained using nitrate is only slightly less than the energy from using oxygen, not just bacteria and archaea but also some fungi and a few other eukaryotic microbes carry out some of the steps in denitrification and can release nitrous oxide.

The end product of denitrification carried out to completion is nitrogen gas, but sometimes the pathway stops at nitrous oxide. Some denitrifying microbes release only nitrous oxide because they are missing a key enzyme, nitrous oxide reductase, that takes nitrous oxide to nitrogen gas. Most denitrifying fungi don't have nitrous oxide reductase and may be a large source of the greenhouse gas in soils where fungi are abundant.[33] Acidic soils which favor fungi tend to emit large amounts of nitrous oxide. Even those microbes with nitrous oxide reductase can release nitrous oxide during denitrification, especially when nitrate levels are high. The most important factor, however, is oxygen. Heterotrophic microbes use nitrate and carry out denitrification, producing nitrous oxide along the way only when oxygen goes to very low levels; how low is an area of much research.

The other main microbial pathway producing nitrous oxide is nitrification. During nitrification, ammonia is oxidized eventually to nitrate by microbes to gain energy. Whereas denitrification depletes nitrate from the environment, nitrification puts it back. Nitrification is an example of chemolithotrophy, a term derived from the Greek for "rock eaters." A few microbes, the chemolithotrophs, can squeeze energy out of rocklike, inorganic chemicals (Box 8.1). Although the term "rock eaters" may be a bit over the top, the phrase highlights how alien chemolithotrophy is and how radically different it is from the well-known way, organic matter degradation, in which animals and many microbes gather energy for growth and reproduction. Whereas animals gain energy from oxidizing organic matter, chemolithotrophs get their energy by oxidizing inorganic compounds like ammonia. Chemolithotrophy is not as prevalent as other types of microbial metabolisms today, but it may have been the only metabolism around during the early days of the Precambrian. The first cell on Earth was likely a chemolithotroph, and if there is life on Mars, it may also be a chemolithotroph.[34] All chemolithotrophs are either bacteria or archaea. No eukaryote uses chemolithotrophy, because it yields too little energy, about 10 times less than what is produced by organic matter degradation, to support the lifestyle of eukaryotes. For that reason, nitrification is carried out only by a few kinds of bacteria and archaea.

The discovery of chemolithotrophy is so important that it is often taken as the start of microbial ecology as a scientific discipline, and the man who first worked on chemolithotrophy, Sergei Winogradsky, is often considered to be the founder of the field.[35] Born in Kiev, Ukraine, on September 1, 1856, Winogradsky grew up in a wealthy, privileged family and attended one of the local gymnasia chosen by his father, a bank director, because it offered Latin and Greek whereas the others had only Latin. After graduation, young Winogradsky studied law for two years at the University of Kiev, then transferred to the university's Division of Natural Sciences but found the classes there boring. At the age of 20, in need of a more radical change, he enrolled in the famed St. Petersburg Academy of Music, to study piano.

Box 8.1 How Microbes Eat Rocks

To understand how rock eaters, or chemolithotrophs, wring energy out of inorganic chemicals like ammonia, we need to follow electrons as they move from one molecule to another. In chemoorganotrophy, more commonly called respiration, the metabolism used by animals and many microbes, electrons move from organic chemicals, the electron donor, to oxygen, the electron acceptor. The chemical equation for chemoorganotrophy can be cut into two halves, one for the electron donor:

$$\text{Organic chemical} \rightarrow CO_2 + \text{electrons}$$

and another for the electron acceptor part of the reaction:

$$O_2 + \text{electrons} \rightarrow H_2O$$

As those electrons move, heterotrophic organisms gain energy. In chemolithotrophy, the organic chemical is replaced by ammonia (NH_3), in the case of nitrification:

$$NH_3 \rightarrow NO_3^- + \text{electrons}$$

which when combined with the electron acceptor equation just given yields:

$$NH_3 + O_2 \rightarrow NO_3^-$$

(With apologies to chemists, the equation is not balanced.) Analogous to respiration, as electrons move from ammonia to oxygen, nitrifying microbes gain energy.

(According to Selman Waksman, the Nobel Prize–winning, Ukrainian-American microbiologist mentioned in Chapter 3, Winogradsky was not only a "profound thinker" but also a "brilliant musician."[36]) Winogradsky lasted only a year at the Academy. He transferred once again, this time to the faculty of natural science in St. Petersburg where he finally found an academic home. For his master of science thesis, he isolated a pathogenic yeast

that attacked sugar beets and caused serious problems for the sugar beet in-
dustry. Although the work was never published, he was invited to stay on for
additional training that would have led to a professorship in St. Petersburg.
But he didn't particularly like the type of microbiology being done in St.
Petersburg, and the Russian climate was too harsh for his new wife, Zinaida
Alexandrovna, to whom he remained devoted until her death 60 years
later. Instead of remaining in St. Petersburg, he moved to the University of
Strasbourg where in 1887 he showed that the bacterium *Beggiatoa* gains en-
ergy from a form of chemolithotrophy based on sulfur.

Winogradsky used his work on *Beggiatoa* as a jumping-off point to study
another example of chemolithotrophy, nitrification.[37] He and others found
that the most common type of nitrification consists of two steps, carried out
by two different types of microbes. The first step, called ammonia oxidation,
takes ammonia to nitrite, and is carried out by bacteria such as *Nitrosomonas*,
first isolated and named by Winogradsky. The second step, carried out by
Nitrobacter and related bacteria, finishes the job by oxidizing nitrite to nitrate. Only the first step, ammonia oxidation, releases nitrous oxide. The
gas can also be produced by another version of nitrification, "comammox"
(complete ammonia oxidization), discovered more than a century after
Winogradsky's work. In comammox, a single bacterium oxidizes ammonia
all the way to nitrate. However, comammox doesn't appear to release much
nitrous oxide.[38] Another group of bacteria that oxidizes ammonium anaerobically without oxygen doesn't produce any nitrous oxide.[39]

In addition to ammonia-oxidizing bacteria, some archaea are
chemolithotrophs and oxidize ammonia to gain energy. The discovery of
ammonia-oxidizing archaea in the early 2000s was big news in part because it
brought together two exotica in microbiology: archaea, a recently discovered
form of life, and chemolithotrophy, an alien way to generate energy. Archaea
were known to be abundant in soils and in the deep ocean where they make
up about 20 percent of all microbes,[40] but what all these archaea do wasn't
clear. They weren't all methanogens, the first microbes to be called archaea
(Chapter 1), because the new, abundant archaea were in well-oxygenated waters and soils where methane isn't produced. The initial clue that archaea can
oxidize ammonia came from one of the first studies to use "metagenomics,"
an approach that retrieves genomic information from several organisms all
at once. The study was led by Craig Venter.

In the early 2000s, Venter had just finished leading a private effort to sequence the human genome when he took his sequencing technology to the
Sargasso Sea.[41] In addition to looking for an excuse to use his sailboat the *SV*

Sorcerer II, he went to the Sargasso to sequence everything he could find. He and his team didn't follow the traditional approach, which would have been to isolate the microbes from seawater, grow them up alone in "pure culture" in the lab, and then sequence each microbe one by one. (Sequencing yields the order of the four base pairs (A, T, C, or G) in the genes making up an organism's genome: the set of instructions for how it makes a living.) One of several problems with the traditional approach is that many microbes in the Sargasso and indeed in all habitats, including the human gut, cannot be isolated and grown in pure culture. Venter's solution was to skip the isolation step and go straight to sequencing. Instead of sequences from a single isolated organism, the metagenomic approach used by Venter returns sequence data potentially from all organisms in the sample. The paper reporting the sequence data from the Sargasso Sea was published in *Science*, one of the most prestigious outlets in science, mainly because Venter's use of metagenomics was audacious, and the amount of sequence data presented in the paper was mind-boggling at the time. Buried near the end of the paper was one sentence about an ammonia-oxidation gene Venter's team found connected to genes from archaea. That was enough to set in motion a frantic search for more. Eventually archaeal ammonia-oxidation genes were found not only in other marine waters but also in soils and just about everywhere nitrification takes place.[42]

Metagenomics remains a powerful tool for looking at nitrification and nitrous oxide production by uncultured microbes in nature, and it's now commonly used to analyze microbial communities in all habitats, from soils to the ocean. It has proven essential in exploring the human microbiome. But a study using the traditional approach was necessary to prove unequivocally that archaea can gain energy by oxidizing ammonia as part of the first step in nitrification. The metabolism of these archaea was confirmed by detailed studies of *Nitrosopumilus maritimus*, which was isolated from the sediments of a marine aquarium.[43] Those studies found that *Nitrosopumilus*—and by implication other ammonia-oxidizing archaea—release much less nitrous oxide than ammonia-oxidizing bacteria.[44] But because they are so abundant, ammonia-oxidizing archaea are big contributors to nitrous oxide emissions.

Competition among ammonia-oxidizing bacteria, ammonia-oxidizing archaea, and comammox bacteria contributes to determining how much nitrous oxide is released from agricultural soil or a natural habitat, but it is difficult to look at what one type of microbe is doing versus another.[45] More is understood about the factors setting overall rates of nitrous oxide release by nitrification. Along with acidity and levels of ammonia and nitrate, the most important factor is oxygen. The crucial role of oxygen is clear in one model of ammonia oxidation in which a key but toxic intermediate is nitric oxide (NO).[46] When oxygen is plentiful, nitric oxide is oxidized to nitrite, but when oxygen levels are low, nitric oxide must be removed by reducing it to nitrous oxide. Overall, a decrease in oxygen levels leads to less ammonia oxidation but a higher relative release of nitrous oxide.

So oxygen is crucial in setting how much nitrous oxide is released by both denitrification and nitrification. In the absence of any oxygen, aerobic ammonia oxidation by all microbes stops and only denitrification can proceed, making it the only source of nitrous oxide. With high levels of oxygen, denitrification is shut down and nitrification takes over. Both pathways are possible in low oxygen. Which pathway wins depends in part on what "low" actually means and on other variables like concentrations of nitrate, ammonium, and organic matter. Overall, denitrification seems to produce the most nitrous oxide. Estimates based on nitrogen isotopes indicated that while nitrification was important in well-oxygenated waters, denitrification was the main mechanism in low oxygen waters in the Eastern Tropical Pacific Ocean.[47] It was also the main mechanism in soils from a corn field and subalpine forest, according to a recent study also using nitrogen isotopes.[48] Further support for denitrification comes from a meta-analysis of 109 studies that examined the effect of nitrogen fertilizer on nitrous oxide emission.[49] The meta-analysis found that, although all forms of fertilizer stimulated nitrous oxide emissions from both natural habitats and agricultural fields, application of nitrate resulted in the highest emissions, about three times higher than was observed for ammonium or urea. Because nitrate is the starting chemical for denitrification, these results point to denitrification as being the main source of nitrous oxide. To sum up, nitrification is the main source of nitrous oxide in well-oxygenated waters and soils, but denitrification is a bigger source when oxygen levels are low or completely absent.

Even when denitrification is the immediate source of nitrous oxide, ammonia-oxidizing bacteria and archaea and other microbes are essential for providing the nitrate needed by denitrification. (It is worthwhile to look at Fig. 8.3 again.) Heterotrophic microbes use oxygen to degrade manure, urea, and other organic matter to ammonia, which then is transformed by ammonia-oxidizing bacteria and archaea to nitrite. Other microbes take over and oxidize nitrite to nitrate, the starting point for denitrification. The oxidation of organic matter, ammonia, and nitrite can deplete oxygen to levels low enough for denitrification to dominate; denitrifying microbes continue to degrade organic matter using nitrate while potentially releasing nitrous oxide. So denitrification depends on nitrification; a microbial ecologist would say the two are coupled. In short, it takes a community of microbes to produce nitrous oxide.

Getting Rid of Nitrous Oxide

The previous section about nitrous oxide sources had to be long to adequately describe the many pathways used by microbes to produce this potent greenhouse gas. This section about nitrous oxide degradation can be short; degradation depends on only a few chemical and microbial reactions. The main mechanism is a photochemical one in the stratosphere where ultraviolet light is intense enough to reduce nitrous oxide to nitrogen gas.[50] In contrast, microbes remove little atmospheric nitrous oxide, about 0.01 Tg N per year, which is next to nothing compared with the stratospheric sink of 13.5 Tg N.[51] Unlike the several enzymes that produce nitrous oxide, only one enzyme, nitrous oxide reductase, is thought to degrade the gas.[52] Also unlike its production, nitrous oxide is degraded only by archaea and bacteria, not by fungi or other eukaryotes. Microbiologists and microbial ecologists know the most about how denitrifying archaea and bacteria convert nitrous oxide to nitrogen gas when levels of nitrate are low and oxygen is absent. In this case, nitrous oxide is an electron acceptor, taking the place of oxygen or nitrate. The gas is a poor replacement, however, yielding less energy than generated when oxygen or nitrate is the electron acceptor.

Going the other way, the oxidization of nitrous oxide to nitrite or nitrate, should be another mechanism to consume nitrous oxide. Although nitrous oxide oxidation wouldn't generate a lot of energy, it's no worse than some other chemolithotrophic pathways. But it isn't used by any microbe,[53] making it one of the rare cases in which microbes appear not to take advantage of a thermodynamically possible reaction to generate energy. Perhaps microbiologists just need more time, patience, and probably a bit of luck to find it. There are a few examples in which microbiologists found a microbe carrying out a particular reaction long after it was predicted to exist. One example is a single microbe carrying out complete nitrification; comammox was conjectured nearly 10 years before microbiologists finally found it.[54] For now, reduction of nitrous oxide during denitrification is the only way microbes remove the gas.

Perhaps only one degradation pathway would be enough if nitrous oxide were more like methane. As discussed in Chapter 7, methane is produced only in oxygen-depleted, or anoxic habitats but can be degraded by bacteria and archaea in both oxygen-rich and oxygen-poor soils and waters. In many environments, oxygen-rich waters, sediments, and soils, filled with methanotrophic bacteria, are an effective barrier between anoxic, methane-rich habitats and the atmosphere. That's not necessarily the case for nitrous oxide. It's produced with and without oxygen, but it seems to be degraded only in the absence of oxygen. As a result, in the oceans, nitrous oxide released from oxygen minimum zones, which are oxygen-poor waters usually about 200 to 1000 meters below the surface, is unlikely to be degraded during its transit through oxygen-rich waters to the surface and the atmosphere. Like the oceans, soils don't have a layer with intense nitrous-oxide degradation that would stop escape of the gas to the atmosphere. Perhaps nitrous oxide degradation is more complicated than now known, but for now, it seems easier for microbes to make this potent greenhouse gas than to degrade it.

The Future

The rise of atmospheric nitrous oxide over the last 75 years can be tied to an increase in nitrogen fertilizer, which was needed by agriculture to feed the human population as it grew from about 1.6 billion in 1900 to 8 billion today. The same dynamics will shape the future of nitrous oxide. Even more fertilizer probably will be needed by agriculture to improve the lives of millions of people now without adequate nutrition and to feed the additional 1.7 billion people projected to be on the planet by 2050. More intensive nitrogen fertilization is needed in sub-Saharan Africa and elsewhere to increase crop yields and reduce malnutrition, even though global nitrogen inputs, mainly in rich countries, have already surpassed the "planetary boundary" for nitrogen by as much as 142 percent.[55] Predicting how much more fertilizer will be needed in the future depends on several factors, such as the amount of land devoted to each crop (corn requires more fertilizer than soybeans), the implementation of practices for maximizing fertilizer use by crop plants (the 4Rs, or using the right fertilizer at the right rate at the right time at the right place), and the nitrogen use efficiency. As mentioned before, the efficiency at which a plant takes up the applied fertilizer nitrogen varies with the type of plant and among countries. Depending on the values assumed for the nitrogen use efficiency and other factors, nitrogen fertilizer use could climb as much as 300 percent by 2050 compared with today. Even a smaller increase in fertilizer use guarantees the release of more nitrous oxide into the atmosphere.

Adding to the impact of nitrogen fertilizer on nitrous oxide is climate change. Global warming has already contributed both directly and indirectly to the rise in nitrous oxide levels.[56] Higher temperatures directly stimulate microbial reactions that produce nitrous oxide and indirectly affect those reactions by reducing oxygen solubility. The oxygen content of soils depends on soil moisture, which varies with temperature and precipitation. Global warming acting alone would dry out soils and lessen nitrous oxide emission, but a bigger driver of soil moisture is precipitation. The wet regions of the world may get even more precipitation and release even more nitrous oxide.

How global warming may change nitrous oxide emissions from the oceans is complicated. Warming waters are expected to increase the size of oxygen minimum zones in the subtropical oceans where nitrous oxide production is most intense. Bigger oxygen minimum zones would mean more nitrous oxide production, yet warmer surface waters causing stronger stratification could lead to less; stratification could slow down the export of organic matter that when degraded releases the starting chemicals, ammonia and nitrate, for nitrous oxide production by nitrification and denitrification, respectively. One modeling study foresees oceanic nitrous oxide emissions declining 24 percent by 2100,[57] while another hypothesizes that nitrous oxide will decline slightly at first but then over two millennia, it will increase by about 21 percent as the oceans continue to lose oxygen.[58] Ocean acidification is another climate-change factor that could result in higher nitrous oxide emissions, perhaps over 490 percent higher by 2100, according to a study in the Subarctic North Pacific Ocean. Although the way nitrous oxide emissions will respond to climate change isn't completely known, it's clear that more climate change means more nitrous oxide. It's another example of what climate scientists call positive feedback, even though it's not a positive for the planet's climate.

So far, it seems that the future for nitrous oxide is more of the same. More fertilizer will mean more nitrous oxide, egged on by global warming, changes in precipitation, and acidification in the ocean. Global warming, however, is also bringing in a wild card: the thawing of permafrost in high-latitude and mountainous ecosystems. As discussed before, climate scientists are very concerned about the release of carbon dioxide and methane from thawing permafrost—a "bomb" waiting to be detonated by microbes as they degrade once frozen soil organic matter. Nitrous oxide may make the bomb even bigger. Today nitrous oxide emissions from high-latitude habitats are low, and boreal and Arctic wetlands can even be net sinks for the gas.[59] (Most studies focus on the north, as not much is known about Antarctica, and the amount of permafrost carbon in ecosystems like the Tibetan Plateau is small compared to what's in northern high latitudes.) Emissions are low not only because of cold temperatures but also because nearly all nitrogen in these habitats is locked up in organic matter and is unavailable to be converted into nitrous oxide. All that may change as global warming continues to transform alpine, subarctic, and Arctic ecosystems.

Ecologists have examined the impact of higher temperatures by artificially warming permafrost soils in little greenhouses.[60] Roughly a meter wide by a half to one meter tall and open at the top, the mini-greenhouses are made with glass walls that slope inward to block the wind and trap enough heat to warm the air by 1 to 4 degrees Celsius. Even this modest warming is enough to stimulate nitrous oxide emissions by 10 to nearly 500 percent, with the exact effect varying with location. Other work suggests that high-latitude habitats that are now net sinks of nitrous oxide could become net sources as temperatures continue to climb. Warming speeds up nitrous oxide production by nitrification and denitrification while it also stimulates the release of more ammonia, the starting point for nitrification, as organic matter previously frozen in permafrost is degraded. But global warming does more than just raise the temperature.

Global warming has transfigured the tundra landscape, leaving behind transformations visible even to the casual observer. How nitrous oxide levels will respond is less clear. Thawing permafrost damages more than just human infrastructure. Along with slumping buildings, buckles in roads and airport runways, and cliffs crumpling into the Arctic Ocean, warming temperatures have caused "collapse features" in the tundra, the thermokarsts mentioned in Chapter 7, which have expanded 60-fold between 1984 and 2015, the last year with good data.[61] Thermokarsts form when land rises by frost heaving as winter sets in and then collapses in the summer as permafrost thaws, leaving behind fens, bogs, or sometimes shallow ponds with unfrozen bottoms, called taliks. One modeling study argued that emission of nitrous oxide increases when thermokarsts first form but then decreases over years as meltwater drains away.[62] Authors of another modeling study were less optimistic and thought emissions would increase as the nitrogen released by warming is more available for nitrous oxide–producing microbes than for plants.[63]

Another visible alteration is the greening of subarctic and Arctic ecosystems; according to satellite data, vegetation has grown higher and denser, and it covers more land as temperatures have risen. Whether greening increases or decreases nitrous oxide emissions depends on how grasses and shrubs affect soil moisture, soil temperature, and nitrogen availability.[64] The projected increase in precipitation (over 50 percent by 2100) is still another climate-change factor that could lead to higher emissions. Barren permafrost lands in the Arctic have released 10 times more nitrous oxide during wet summers than during dry ones. But an alpine meadow took up nitrous oxide during a rainfall only to return to emitting the gas days later when the rain stopped. Perhaps more so than for carbon dioxide and methane, the future of nitrous oxide is murky and full of unknowns. Scientists always say they need more data, but it's especially true in this case.

For now, climate scientists are not as concerned about nitrous oxide as they are about carbon dioxide and methane being emitted from thawing permafrost and a warming Arctic. The latest report from the Intergovernmental Panel on Climate Change (IPCC) concludes that nitrous oxide emissions from the Arctic will not increase, and an increase is not in climate change models.[65] But it's premature to forget about nitrous oxide in the Arctic and certainly in other ecosystems. Although authors of the IPCC report say they have "medium confidence" that climate change will lead to higher nitrous oxide emissions, they have "low confidence" about the size of the increase. Microbial ecologists have a pretty good idea about how temperature, soil moisture, nitrogen availability, and other environmental properties regulate nitrification and denitrification—the pathways that make and degrade nitrous oxide. The individual parts of the puzzle are clear. What's less clear is how the parts fit together in setting the future of this potent greenhouse gas.

However the parts go together, microbes will be at the center of most things happening with nitrous oxide. It's connected with many residents of the microbial world, from archaea and bacteria to fungi and protists, some that produce the gas and too few that degrade it. Because of its role in the nitrogen cycle, not to mention its appearance in Hollywood movies and laughing gas parties, nitrous oxide would be fascinating enough to attract the attention of microbial ecologists and biogeochemists even if it weren't a potent greenhouse gas. But of course, it is. As if carbon dioxide and methane weren't enough, nitrous oxide is another reason to look at microbes as we grapple with understanding climate change.

9

Microbial Solutions

For the past 10,000 years, ever since the last ice age ended, release of carbon dioxide was matched by consumption, and the global carbon cycle was more or less in balance. Microbes were a big contributor to maintaining the balance and to keeping Earth's climate fairly steady for thousands of years. That balance has been upset, however, by the atmospheric carbon dioxide coming mainly from the burning of fossil fuels. As a result, heatwaves are becoming more common, droughts are longer, hurricanes are more frequent and severe, and sea levels are rising.[1] Earth's climate is changing. To head off further climate change, we need to reduce if not completely cease using fossil fuels. Yet even as we switch to solar panels, wind turbines, and other "green" ways of generating energy, we are unlikely to be rid of fossil fuels completely nor stop other enterprises that emit carbon dioxide. We need to do more. Adding to the challenge are the other greenhouse gases. Decarbonization and green energy wouldn't solve the problems caused by methane from natural sources, nitrous oxide, and other greenhouse gases, which are less dependent on fossil fuels than carbon dioxide is.

Microbes can be part of the solution. Just as they are the unseen agents that drive fluxes of carbon dioxide, methane, and nitrous oxide, microbes can help to stop the rise in those greenhouse gases. The possible solutions with microbial connections to be discussed here fall into three categories. We'll go over efforts to reduce the release of methane and nitrous oxide, and then move on to strategies to take carbon dioxide out of the atmosphere. First, let's look at how microbes help keep fossil fuels in the ground by making biofuels and other types of green energy.

Green Energy

There is little that is green in *Mad Max Beyond Thunderdome*, a dystopic action film set in postapocalyptic Australia, starring Mel Gibson as Max Rockatansky and Tina Turner as Aunty Entity. The movie begins with a panoramic view of a barren wasteland, devoid of vegetation and any sign of civilization. A road twists along the featureless landscape, empty except for dust thrown up by Max's truck, which looks like a covered wagon from a Hollywood western. Rather than horses, much less a gas-powered engine, this truck is pulled by six camels. Suddenly Max is jolted to the ground by a projectile that is part toilet plunger, part fly swatter, shot by a thief and his young son flying a decrepit airplane. The thief drops into the truck and takes off before Max can catch him. After pulling on his boots, Max follows the track left by his stolen truck to a trading outpost, Bartertown. There he meets its founder and ruler, Aunty Entity, and argues with her for the return of his truck and provisions. She agrees to help him if he fights MasterBlaster, who controls the town's energy source: methane made from pig manure produced in a subterranean refinery. Being renewable, the methane is about the only green thing in the movie.

In real life as in the movies, manure from pigs as well as from other livestock along with plant debris can be turned into methane, often called biomethane to distinguish it from fossil fuel methane. Manure and plant wastes can be fed into anaerobic reactors where a complex microbial community turns the organic matter into biogas, which is a mixture of methane (45–70 percent) and carbon dioxide (30–55 percent), with smaller amounts of other gases.[2] The biogas digester may be on a farm or at a larger, stand-alone facility that takes in wastes from several sources. There are two more places where waste can be transformed into valuable fuel: landfills and wastewater treatment plants. After about a year of use while the refuse piles up, sealing off table scraps and other organic wastes from contact with the atmosphere, landfills become devoid of oxygen below a thin superficial layer: ideal for producing methane-rich biogas. A gas collection and control system for trapping biogas is required by the Clean Air Act for large landfills in the United States.[3] At wastewater treatment plants, biogas is produced near the end of the clean-up process. Sewage organic waste is first degraded mostly by aerobic bacteria using dissolved oxygen, leaving behind clean water and sludge, which is a dark slurry of bacterial cells, their extracellular slime, and undegraded wastes. One way to deal with sludge is to add it to an anaerobic digester where biogas is produced. For some applications, biogas needs to be cleaned up to yield just biomethane.

Getting into some of the microbiological details helps to explain the challenges in making biomethane. Archaea are the only microbes that make biomethane, yet methane production is more complicated than just one reaction. Methanogenesis is the last step among several mediated by other microbes in the anaerobic food chain, a series of linked metabolisms that decompose organic matter in the absence of oxygen (Fig. 9.1). In this version of a food chain, one organism doesn't eat another but rather the end product released by one microbe is passed on to another. As mentioned in Chapter 7, methanogenic archaea can use only hydrogen gas and a couple of carbon sources, most notably acetate, which is only a proton (H^+) away from acetic acid, the main organic chemical in vinegar. Hydrogen gas and acetate are far from the organic matter that makes up manure, the organic refuse dumped into landfills, or the sewage flowing into a wastewater treatment plant. To illustrate the path from organic wastes to methane, let's look at how methane is made from one organic chemical, cellulose, which is a major constituent of crop residues, wood, and paper, including the version that is flushed down toilets and ends up in wastewater treatment plants. Being a polymer of hundreds to thousands of glucose molecules linked together, cellulose must be broken down, or hydrolyzed eventually to single glucose molecules. When oxygen is abundant, glucose is decomposed by aerobic bacteria to carbon dioxide, and that's it; methane isn't produced. In the absence of oxygen, however, several other metabolic pathways kick in, including fermentation. Fermenting microbes, mainly bacteria and yeasts, gain some energy by transforming glucose to acetate, hydrogen gas, and other chemicals.

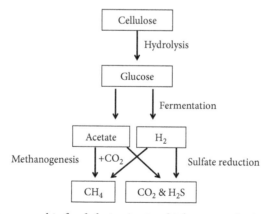

Figure 9.1 The anaerobic food chain. A microbial community is needed to decompose organic matter such as cellulose, which is a polymer of glucose molecules, to methane (CH_4), carbon dioxide (CO_2), and if enough sulfate is present, hydrogen sulfide (H_2S). Key intermediates are acetate and hydrogen gas (H_2).

If any sulfate is present, the acetate and hydrogen gas from fermentation are snapped up by sulfate-reducing bacteria, releasing hydrogen sulfide (Chapter 6), and no methane is produced. In anaerobic digesters, anaerobic reactors, and landfills, however, sulfate is low and sulfate-reducing bacteria are not active enough to prevent methanogenic archaea from using acetate and hydrogen gas to make biomethane.

The problem is that waste organic matter is much more complex than cellulose, and some organic detritus is better than others for producing biomethane.[4] For example, the yield of biomethane from the woody stems of grape vines, composed of hard-to-decompose lignin, is about seven times lower than the yield from wine-making residues, which are rich in carbohydrates that can be directly used by fermenting microbes. Nitrogen-rich manure turns out 30 times less biomethane than cereal grains do. To improve yields, different types of organic wastes may be mixed to lower the overall nitrogen and lipid content, and the starting organic wastes can be pretreated with heat, acid or alkaline digestion, ultrasonication, or enzymes.[5] Pretreatments break up waste organic matter and open it up for attack by the microbes that start the anaerobic food chain, releasing hydrogen gas and acetate that are used by methanogenic archaea to make methane.

Biomethane from biogas digesters has not taken the place of much natural gas, yet it is still valuable where it is commonly used. The biomethane made in the European Union, the world's leader in biogas production, accounts for 4.5 percent of natural gas consumption in Europe.[6] In the United States, which is a distant second in the world ranking, biomethane is equivalent to about 1.4 percent of natural gas use and contributes less than 1 percent of total energy consumption in the country.[7] The United States has much untapped potential to produce more. It has far fewer biogas digesters than the European Union does (1,240 in the United States versus 15,000 in the European Union in 2017), and of the roughly 15,000 wastewater treatment plants in the United States, only about 1,200 harvest biomethane.[8] Capacity has not grown in the United States in recent years; biomethane use was highest in 2015 and by 2022 it had declined 17 percent.[9] Biogas is more valued in rural regions of low-income countries without connections to a centralized power grid or access to cheap natural gas. Even small biogas digesters could provide energy for a farm household while processing farm wastes that otherwise would be a source of atmospheric methane. Small biogas digesters are common in China; the country has 41.9 million small, household digesters providing energy for 200 million people.[10] In short, biomethane is already a success in many parts of the world and could be a bigger energy source in countries like the United States.

There's another way to use microbes to wring energy out of organic debris, this time turning waste into electricity. A history of this electricity-generating method could begin in the late eighteenth century when Luigi Galvani found that a dead frog leg would twitch when subjected to an electrical spark. Galvani's work was followed up by Alessandro Volta and Humphry Davy, whom we've met already in discussing methane and nitrous oxide, respectively, although neither did anything with frogs or body parts from other animals. It wasn't until 1911 that M. C. Potter, a professor of botany in the University of Durham, wondered if a version of Galvani's "animal electricity" was in microbes.[11] After pointing out that energy is released during the decomposition of organic material, he asked if electricity also could be generated by the "vital activity of living organisms."[12] To find out, Potter fed a concentrated glucose solution to microbes in a porous cylinder attached to platinum electrodes and measured the electrical output with a galvanometer. He found a yeast species and one of three bacterial species generated about 0.32 volts over seven minutes. Potter had built what we now call a microbial fuel cell. Rather than yeast or a single bacterial species, today's microbial fuel cells rely on a mixed community of bacteria dominated by *Geobacter*.[13] But the basic idea is the same.

A microbial fuel cell is a battery made of bacteria fed organic matter that produces power from the electrons released as the organics are degraded (Fig. 9.2). In other chapters, I've talked about how microbes gain energy from the cascade of electrons set in motion during the oxidation of organic chemicals, such as glucose, or of inorganic ones such as ammonia in the case of nitrification (Chapter 8). The electrons from these chemicals, the electron donors, pass through a respiratory chain within the microbe and eventually end with an electron acceptor, such as oxygen or sulfate. The microbial fuel cell generates electricity by stealing some of the energy released as electrons cascade down to the electron acceptor.[14] Generation of an electrical current starts on the anodic side of the fuel cell where bacteria release electrons and protons (H^+) as they oxidize organic material to gain energy and then pass electrons to the electrode, while simultaneously protons diffuse over to the cathode. Although some bacteria rely on a soluble chemical to transfer electrons, other bacteria such as *Geobacter* attach directly to the electrode and transfer electrons via proteinaceous protuberances, such as pili or nanowires that act as bridges between the bacterium and the electrode.

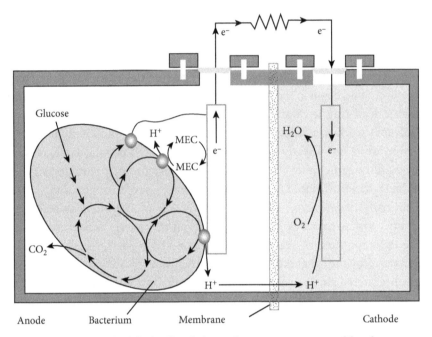

Figure 9.2 A microbial fuel cell. While oxidizing organic matter like glucose, bacteria generate electrons (e⁻), which are transferred to the electrode by either soluble chemicals (mobile electron carrier or MEC) or by direct contact with the electrode, represented here by the long thread connecting the bacterium to the electrode. Protons (H⁺) also released during organic matter oxidation pass through the membrane to the cathode where oxygen (O_2) is reduced to water. Based on Logan and colleagues (2006).

After electrical energy is extracted, the electrons eventually make their way to the cathodic side of the fuel cell where they combine with oxygen and hydrogen atoms to form water. The anode and cathode sides are separated by a permeable membrane that allows protons to pass and the electrical charge to be balanced. As these electrons and protons pass from one side of the fuel cell to another, electricity is generated.

Microbial fuel cells are moving into the real world.[15] They have provided power to a Texas Instruments Chronos digital wristwatch, a smart phone, and a meteorological buoy. The biggest application has been in treating human waste. Here the goal is not only to generate energy but also to decompose organic sewage. Supported in part by the Bill & Melinda Gates Foundation, engineers in Bristol, England, designed Pee Power Urinals for refugee camps that don't have other power sources and then tested the urinals in 2016 at the Glastonbury Festival of Contemporary Performing Arts, an annual four-day musical event in Somerset, England, which attracts about 1,000 people each day.[16] In addition to providing electricity for some of the festival's lighting, 432 microbial fuel cells degraded about 30 percent of oxygen-depleting organic matter (more precisely, chemical oxygen demand) deposited by music-lovers in portable toilets. In another test of Pee Power Urinals, microbial fuel cells were over 95 percent efficient in removing oxygen-depleting organics. A bigger challenge is to deal with the sewage deposited by an entire city into a wastewater treatment plant. A pilot-scale test of this application took place from September 2020 through April 2021 at the Tobyhanna Army Depot in northeastern Pennsylvania.[17] As with the Pee Power Urinals, the test consisted of looking at how much electricity was generated by a microbial fuel cell fed wastewater and how efficient it was in removing oxygen-depleting organic matter either alone or coupled with biofiltration; in the latter case, effluent from the microbial fuel cell was fed into a bacterial community attached to activated charcoal. The microbial fuel cell alone removed almost half of oxygen-depleting organics in the wastewater and over 90 percent when combined with biofiltration while using less than half of the energy required by conventional treatment. If this application is scaled up and becomes common, it could make a significant contribution to reducing fossil fuel use. Moving and treating wastewater consumes about 4 percent of all electricity in the United States.[18]

Biofuels and Corn Ethanol

Perhaps even more alluring than converting waste into biomethane or electricity is using microbes to convert sunlight energy into chemical energy in the form of biofuels. As with most aspects of climate change science, here "bio" mostly means "microbial." The relevant microbes are algae, usually referred to as microalgae in the biofuel literature to distinguish them from macroalgae like kelp. As discussed in earlier chapters, algae and cyanobacteria are the plant-like microbes that use light energy and carbon dioxide to make organic chemicals, like what plants do on land. Cellular organic chemicals from algae such as carbohydrates can be the starting material, or feedstock for making biomethane by the archaea-based approach described before. More intriguing are those algae that have lots of lipids, the oil-like molecules making up membranes, that can be turned into green diesel, or biodiesel, green gasoline, or even green jet fuel.[19] Most biodiesel synthesized commercially today comes from plants, such as rapeseed (canola) and sunflower in Europe, soybean in the United States, and palms in tropical countries.[20] However, algae have several advantages over land plants. Algae grow fast and potentially produce more oil-like chemicals in a hectare than a plant could. The algae can be grown in ponds, tanks, or sunlit bioreactors on nonarable land where crop plant growth would be sparse at best. Depending on the species, algae can use water ranging in salinity from fresh to full-strength seawater that would kill most plants. Algae can also be grown in wastewater and take advantage of its nitrogen and phosphorus nutrients.

Despite these advantages and a long history of applied research, starting with German scientists in the early years of World War II,[21] algal biofuels are still years away from being produced on a large, commercial scale. In the 2000s, several start-up companies tried, but by 2012, they either went out of business or pivoted to producing low-volume/high-value chemicals such as omega-3 fatty acids, astaxanthin (a red pigment found in salmon), and other nutraceuticals (a portmanteau of "nutrition" and "pharmaceuticals").[22] Traditional oil companies like Chevron and Shell have also given up, with ExxonMobil holding on the longest. A website for the fossil fuel giant says it invested about $250 million over the last decade in biofuel research (the company made $19.7 billion in the third quarter of 2022 alone) as part of its efforts to "to advance innovative solutions for a lower-emission energy future."[23] The research has included a collaboration with Synthetic Genomics (now called Viridos[24]) founded by Craig Venter, the genome biologist who was among the first to sequence the human genome and the microbial community of the Sargasso Sea (Chapter 8). The Synthetic Genomics-ExxonMobil team used a new technique in genetic engineering, CRISPR, to edit a gene that resulted in a doubling of an alga's lipid content,[25] making the genetically engineered microbe an ideal candidate for producing biofuels. However, despite the progress, ExxonMobil stopped its support in December 2022, forcing Viridos to lay off 60 percent of its staff.[26] Viridos is soldiering on, however, with money from Bill Gates's Breakthrough Energy and other backers.

Although green fuels from algae are still in the distant future, another fuel made by microbes is very much in the present. The fuel is ethanol—the alcohol in beer, wine, and whiskey. Ethanol added to gasoline can be used in motors without modifications when it's 10 percent of the fuel mixture or less, whereas only flexible-fuel vehicles can use fuel that is 85 percent ethanol.[27] Whether as a fuel or in alcoholic beverages, ethanol is made by microbes carrying out fermentation. Although many bacteria are capable of carrying out ethanol fermentation, the ethanol for fuel and alcoholic beverages comes from natural and genetically engineered strains of one yeast species, *Saccharomyces cerevisiae*.[28]

Using ethanol as a fuel is controversial, in part because of the starting organic matter, the feedstock, used by the fermentation industry. It varies from country to country, depending on what is cheapest, most available, and easiest to turn into the sugars favored by *Saccharomyces*. Wheat is used in some European countries (Belgium, France, and Germany), while sugar beets are the feedstock in others (Austria, Belgium, and the Czech Republic).[29] Bioethanol has been made from coconuts in the Philippines[30] and from sweet sorghum and cassava in China.[31] However, the amount of ethanol made with these feedstocks and all others is trivial compared to the two biggest. Over 80 percent of the ethanol produced in the world is from two plants, corn (maize) and sugarcane, grown in two countries, the United States and Brazil, respectively.[32] (Maize is also used in Europe and China, and sugarcane is a common feedstock in India and tropical countries.)

Thanks to the Corn Belt, the United States is the world leader in growing corn and in producing bioethanol, accounting for about half of all ethanol made globally. Brazil grows more sugarcane than any other country and is second in making ethanol, producing a bit more than 30 percent of the world total. Brazil is second in the bioethanol world ranking, even though it's easier to go from sugarcane to ethanol than from corn.[33] The concentration of sucrose (aka table sugar), the sugar that can be directly transformed by the yeast *Saccharomyces* into ethanol, is high in the juice crushed out of sugarcane. Using the enzyme invertase, *Saccharomyces* cleaves sucrose into its two chemical components, glucose and fructose, and then ferments them to yield ethanol. Additional steps are needed to make ethanol from corn—more specifically, corn kernels, the part of the plant most used in industrial ethanol fermentation. Although sweet corn has a lot of sucrose, kernels of the corn variety used for industrial ethanol fermentation contain mostly starch, which is a polymer of glucose. Unlike many organisms, including humans, natural strains of *Saccharomyces* cannot cleave starch, so it must be broken down using heat, enzymes from other microbes, and a bit of acidity to generate glucose before the yeast can carry out fermentation and produce ethanol. Because of these additional steps, making corn ethanol is 50 to 60 percent more expensive than making sugarcane ethanol.[34] Even so, the additional steps and costs are not the main reasons why corn ethanol is controversial.

The controversy is whether corn ethanol is really a green fuel. Though renewable, it may be no better than the gasoline it's supposed to replace. The problems with corn ethanol were documented by a study published in 2022 that did a cost-benefit analysis of the Renewable Fuel Standard (RFS) first passed in 2005 by the U.S. Congress to increase the use of green fuel for transportation.[35] Today, because of RFS, most gasoline in the United States contains 10 percent ethanol. Although the law says that the ethanol should be made with other plant material, in fact it has been far easier to use corn. Ethanol fermentation facilities now consume about a third of the entire corn harvest in the United States. Because of RFS and its extension in 2007, from 2008 to 2016 cropland devoted to corn expanded 8.7 percent overall in the country and 50 to 100 percent in regions like North and South Dakota and western Minnesota. RFS also reduced the acreage that would have been converted back to natural grasslands if the law wasn't around. Even though the Conservation Reserve Program pays farmers for the conversion, more money could be made by growing corn. Because of greater corn cultivation, nitrogen fertilizer increased by 7.5 percent from 2008 to 2016, according to the 2022 study, and predictably, with more nitrogen fertilizer, nitrous oxide release also increased by 8.3 percent after RFS was enacted. Other environmental problems tied to RFS include more soil erosion and greater nutrient pollution of waterways. The 2022 study concluded that in terms of greenhouse gas emissions and other environmental costs, corn ethanol is no better than gasoline and may in fact be worse.

Another argument against both corn and sugarcane is that using either to make ethanol diverts food for people to fuel tanks of cars and light trucks. The same criticism applies to many other feedstocks, such as sugar beets, wheat, or sweet sorghum used for ethanol fermentation. The problem is solved by using plant parts rich in lignocellulose that aren't edible by humans.[36] "Ligno" refers to the main chemical, lignin, giving wood its strength (see Chapter 3), and cellulose is the main building block that plants and many algae use in their cell walls. Plant debris with lots of lignocellulose is cheap and readily available, but it is difficult to convert to the starting chemicals, mostly sugars like glucose that yeast can ferment to ethanol. We've seen already that lignin is difficult for microbes to degrade. Cellulose is easier, although not when linked to lignin. Research is currently focused on finding the right combination of plant wastes (lignocellulose differs among plant species), heat and chemical pretreatment, and microbes better suited to carry out ethanol fermentation with the byproducts released by lignocellulose decomposition. Genetically engineered yeasts may be needed to ferment more sugars than just the sucrose favored by *Saccharomyces*, including hexoses (six-carbon sugars) other than glucose and pentoses (five-carbon sugars). In the meantime, production of second-generation bioethanol from lignocellulose wastes isn't economically competitive with first-generation bioethanol from corn or sugarcane, not to mention fossil fuel gasoline at current prices.

Whatever the feedstock, bioethanol and other microbial fuels add up to a lot of energy. These renewable sources haven't been publicized in the American popular media like wind turbines and solar panels have been, yet until recently microbes made more energy in the United States than what turbines or solar panels cranked out (Fig. 9.3). Microbial energy was ahead of wind power until 2020 and still contributes more energy than solar panels do. Recent headlines about the huge increases in solar and wind power are about new investments, not facilities already existing. Microbes made about 23 percent of all renewable energy used in the United States in 2022.[37] Now we just saw that some of that microbial energy may not be so green, although wind turbines and solar panels also come with costs to the environment. These environmental costs can't be ignored, yet the amount of energy

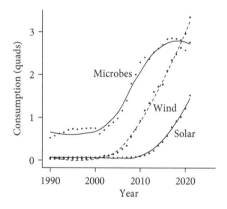

Figure 9.3 Consumption of energy generated by microbes, wind turbines, and solar panels in the United States. Microbial energy is the sum of energy from "wastes" (wastewater treatment plants, landfills, and biogas digesters) and biofuels (bioethanol, biodiesel, and renewable diesel). Consumption is measured in quadrillion British thermal units (quads). Data from U.S. Energy Information Administration (https://www.eia.gov/).

generated by microbes is still impressive. If the costs can be reduced, microbial energy could contribute even more to stopping the rise in greenhouse gases. Mad Max's world ran on only pig manure, but in the real world, we should take advantage of all forms of renewable energy, including the many ways generated by microbes.

Stopping Cow Belches and Laughing Gas

Renewable energy from microbes and other sources replaces fossil fuels and reduces emissions of the number-one greenhouse gas, carbon dioxide. Reducing the extraction, transportation, and burning of coal, oil, and natural gas would also help lower methane emissions, but recall that most methane comes from other sources (Chapter 7). One of the biggest is agriculture, which accounts for nearly 40 percent of methane emitted by human activity.[38] Agriculture is also responsible for about half of all nitrous oxide released into the atmosphere (Chapter 8). Because emission rates of methane and nitrous oxide are less than that of carbon dioxide, the contribution by agriculture to the total greenhouse gas impact is lower yet still substantial. Production of the food needed to feed eight billion people accounts for about a third of all greenhouse gas emissions.[39] Reducing emissions from agriculture would be a big step in slowing the rise of especially methane and nitrous oxide in the atmosphere.

We'll start with the supposedly low hanging fruit of greenhouse gases, methane. It seems like it should be easy to greatly reduce methane from agriculture because most of this potent greenhouse gas comes from just one source: livestock.[40] Beef cattle, the coal power plants of agriculture, release almost 40 percent of the agricultural total.[41] Next are dairy cattle, which account for 17 percent, while other ruminants such as sheep and goats contribute less than 10 percent. Pigs and chickens, which are not ruminants, don't produce much methane, although their manure can. The obvious solution to reducing agricultural methane is to get rid of at least beef cattle if not also dairy cows. That has happened to some extent in high-income countries like the United States where people are eating more white meat and somewhat less beef; from the 1960s to 2010s production of chicken and pork in the United States increased by about 500 and 100 percent, respectively, while beef production went up by only 34 percent,[42] even though the country's human population grew about 80 percent. Yet methane from livestock has not slowed down. New hot spots of beef consumption and methane emissions have emerged in South Asia (34.7 teragrams or Tg of methane per year in the 2010s), Brazil (12.0 Tg), and China (11.0 Tg), all releasing more methane than the United States (8.3 Tg) or western Europe (7.2 Tg).

The obvious solution, having fewer cattle, may be as easy to implement as getting rid of coal power plants: possible but hard. Fewer cows would be needed if we got our milk from yeast and our hamburgers from plants.[43] Unless microbial and plant alternatives to milk and beef are embraced by many more people than is the case today and beef consumption decreases dramatically, there is still the need for other strategies to limit methane coming from cows.

Some methane, about 10 percent of livestock emissions, is released during the decomposition of the manure excreted by these animals. The manure could be fed into a biogas digester as discussed before and the biomethane captured to power farm operations, or the manure could be applied onto fields as fertilizer (but see later). Other countermeasures are needed for manure when it is stored. Some studies have looked at dosing it with sulfuric acid (H_2SO_4).[44] Although the drop in pH may help, adding the sulfate (SO_4^{2-}) in sulfuric acid would stimulate sulfate-reducing bacteria which outcompete methanogenic archaea for their favored energy sources. The result is less methane. That hypothesis explains why adding potassium sulfate, which doesn't increase acidity, also reduces methane production. Another effective addition to manure is biochar, which is the charcoal-like residue leftover from burning wood and other plant material without much air. According to a study done at the Philip Verwey Dairy in Madera, California, composting manure with biochar reduced methane emissions by nearly 80 percent.[45] The biochar allows more air to penetrate the manure by soaking up water and creating more air-filled pores. More oxygen means less methane.

Sulfate or biochar wouldn't help with the main route by which cattle and other ruminants release methane: belches with smaller amounts released by flatulence. Between the two exits from the cow is the rumen, a stomach-like sac where methane is produced. It is also where a version of the anaerobic food chain decomposes plant debris and releases organic acids like acetate, propionate, and butyrate, which the animal takes up for its subsistence. Without the rumen microbiome, a ruminant would starve to death regardless of how much grass or grains it eats. The rumen microbiome is large and complex, packed with bacteria (10^{10}–10^{11} per gram of rumen fluid), protozoa (a million per gram), and many methanogenic archaea.[46] Researchers have looked into a variety of weapons to attack those archaea, including vaccines, chemicals that inhibit methane-producing enzymes, and extracts of red macroalgae, or seaweeds.[47] The seaweed idea is being pursued by several companies, including CH4 Global, Sea Forest, and Rumin8 in Australia and Symbrosia and Blue Ocean Barns in Hawai'i.[48]

Another agricultural source of methane is rice cultivation.[49] Rice contributes roughly the same percentage to global emissions, about 10 percent, as what manure releases, but more effective strategies are available to reduce emissions from rice fields. One is to minimize flooding of rice fields and shrink the low-oxygen habitats that favor methanogenic archaea. Using water management to lower methane emissions without sacrificing yields would be helped by growing drought-tolerant varieties of rice. Another strategy is to add biochar to rice fields, like the biochar treatment of manure. Although biochar may change oxygen levels and other soil properties that work against methanogens, it also appears to stimulate methanotrophic bacteria and methane degradation.[50] Whatever the mechanism, it is effective. Biochar can reduce methane emissions from rice fields by over 80 percent.[51]

When it comes to nitrous oxide, the equivalent of cows or coal power plants is nitrogen fertilizer. The huge expansion and intensification of agriculture and its reliance on fertilizer are big reasons why levels of laughing gas in the atmosphere have been increasing over the last 50 years (Chapter 8). The obvious solution is to use less fertilizer. That may be even harder to do than getting rid of cows or coal power plants. From the perspective of the farmer wishing to maximize crop yields, it's better to err on the side of adding too much fertilizer than too little. As with cows and coal power plants, changing a farmer's attitude doesn't have much to do with microbes, yet a bit of microbiology helps to explain some current farming practices, their consequences, and some solutions.

One decision facing farmers is about one of the 4Rs mentioned in Chapter 8 for effective fertilization: what is the right fertilizer? In terms of nitrogen, the choices are anhydrous ammonia, ammonium, nitrate, urea, or manure. When it comes to nitrous oxide emissions all forms are about equally bad, except manure is the worst.[52] Manure appears to supply both the nitrogen and the organic carbon needed by denitrifying bacteria to turn nitrate into nitrous oxide. Even though it's renewable and "organic," manure may not be so "green" if handled improperly and applied to fields in excess of what crop plants need.

Farmers don't want expensive and valuable fertilizer nitrogen to disappear into air, so they use chemicals to inhibit nitrification, a problem with ammonia and ammonium fertilizers, and to slow down urea degradation. Nitrapyrin is one of several nitrification inhibitors on the market while N-(n-butyl) thiophosphoric triamide stops the enzyme urease from cleaving urea into carbon dioxide and ammonium. Unfortunately, these chemicals don't seem to have much impact on nitrous oxide unless they are used along with reducing the amount of nitrogen fertilizer applied to a field.[53] Rather than a broad attack on nitrification, a more subtle approach is to use a chemical like 3,4-dimethylpyrazole phosphate that targets only ammonia-oxidizing bacteria, not their archaeal counterparts, in the first step of nitrification.[54] Because ammonia-oxidizing archaea release less nitrous oxide (Chapter 8), anything favoring them over ammonia-oxidizing bacteria should lower nitrous oxide emissions. Along with chemicals that select for ammonia-oxidizing archaea, the same result has been achieved with a bacterium. In an experiment with rapeseed, addition of the bacterium *Bacillus amyloliquefaciens* increased soil acidity, which favored ammonia-oxidizing archaea, and lowered nitrous oxide emissions by over 40 percent compared to a control without the added bacterium.

Another strategy may help lower emissions of nitrous oxide after it's made. Unlike the many ways of making the potent greenhouse gas, recall that only a single enzyme, nitrous oxide reductase, can degrade it by reducing it to nitrogen gas (Chapter 8). The strategy is based on the bacterium *Bradyrhizobium japonicum*, already used in agriculture as a biofertilizer.[55] The bacterium has the gene for nitrous oxide reductase, *nosZ*. Agricultural microbiologists in Japan tested a natural strain of *B. japonicum* with nitrous oxide reductase activity (*nosZ+*) and mutant strains with even higher activity (*nosZ++*) to see if they could lower emissions from soybean plants.[56] In laboratory experiments, nitrous oxide degradation by *nosZ++* strains was much higher than degradation by the natural *nosZ+* strain, not to mention the strain without any nitrous oxide reductase (*nosZ-*), and nitrous oxide emission was as much as 90 percent lower. The *nosZ++* strategy is the last line of defense stopping nitrous oxide. The second-to-last defense is to inhibit the conversion of fertilizer nitrogen to nitrous oxide by nitrifying and denitrifying bacteria. The most effective defense in minimizing nitrous oxide emissions from agriculture, however, is the first one: use less nitrogen fertilizer.

Farmers may be able to get by with less nitrogen fertilizer, at least the fixed nitrogen made by the Haber-Bosch process, if their crop plants can take up nitrogen made by bacteria like *B. japonicum*. It's called a biofertilizer because it's a nitrogen-fixing bacterium that supplies nitrogen nutrients to crops without help from the Haber-Bosch process. As mentioned before, nitrogen-fixing bacteria turn nitrogen gas into fixed nitrogen, here ammonium, that can be used by plants and other organisms. Although many nitrogen fixers grow in soils and aquatic habitats without close contacts with plants, the nitrogen-fixing bacteria most important in agriculture live symbiotically in the roots of legumes like soybeans, peas, and lentils. The problem is, corn is not a legume and doesn't have symbiotic nitrogen-fixing bacteria, so it requires high amounts of nitrogen fertilizer. One solution is to make corn more like soybean. Addition of a nitrogen-fixing bacterium to corn fields has increased yields and lowered the need for nitrogen fertilizer.[57] Although the bacterium colonizes corn roots, it does not appear to enter an intricate symbiotic relationship within corn roots like that seen between soybean and its symbiotic nitrogen-fixing bacteria. In any case, similar experiments with adding nitrogen-fixing bacteria to wheat and rice have been successful, and the idea has been commercialized by a biotech firm in California.[58]

Adding nitrogen-fixing bacteria to crop plants is an example of "environmental microbiome engineering," which can be defined as the alteration of a natural microbial community to reach a practical goal, such as promoting crop plant growth.[59] Although it's a new term, coined to highlight the use of genomic science in applied environmental microbiology, altering natural microbial communities has been around at least since 1896 when a patent was taken out in the United States for "Nitragin," which consisted of nitrogen-fixing bacteria mixed with gelatin.[60] It was used as a nitrogen fertilizer before the invention of the Haber-Bosch process when farmers had few other options. Nitragin is still sold today to stimulate the growth of alfalfa and sweet clover,[61] even though those two legumes could acquire symbiotic nitrogen-fixing bacteria without human intervention. (Because of nitrogen fertilization and domestication, cultivated legumes get less nitrogen from their symbiotic nitrogen-fixing bacteria than their wild ancestors do.[62]) Nitrogen-fixing bacteria mixtures are also applied to another legume, soybean, especially in Brazil.[63] Environmental microbiome engineering potentially could be used to combat climate change in other ways, such as to enhance the formation of soil organic matter.

What about nitrous oxide? It's not clear if emissions would be lower when the nitrogen is from a nitrogen-fixing bacterium rather than from the Haber-Bosch process. The real problem is crop plants like corn that, without environmental microbiome engineering, require high amounts of nitrogen fertilizer. We can hope that those crops supplemented with nitrogen-fixing bacteria would become more like soybeans in another aspect: soybean with its symbiotic nitrogen fixers release much less nitrous oxide than does corn fed a traditional nitrogen fertilizer.[64] Harnessing nitrogen fixers is one of potentially many ways in which microbes can help to solve the nitrous oxide problem.

Negative Emissions

Our to-do list for stemming the rise in greenhouse gases is already quite long, from switching to renewable energy made by microbes to using nitrogen-fixing bacteria as biofertilizers. Unfortunately, more action is needed if we are to head off even more global warming than we are now experiencing. In thinking about the future of the planet's climate, scientists working on the Intergovernmental Panel on Climate Change (IPCC) have come up with thousands of scenarios that make different assumptions about "shared socio-economic pathways" and the climatic response.[65] The upshot of all the sophisticated crystal-ball gazing is that more needs to be done than just striving to eliminate fossil fuel use.[66] All scenarios with a chance of keeping global warming to less than 1.5 degrees Celsius include efforts to remove 100 to 1000 petagrams of carbon dioxide from the atmosphere by 2100, a big fraction of the nearly 700 petagrams humans have released since the nineteenth century.[67] In short, we need "negative emissions technologies," to use climate science jargon, to take carbon dioxide out of the atmosphere or to augment the consumption mechanisms already operating as part of the natural carbon cycle. Microbes can be part of the first and are already helping with the latter.

One approach is to capture carbon dioxide before it gets into the atmosphere. One of the first scientists to think about the problem, Cesare Marchetti, jested in a paper published in 1977 that

> This is not the place to think of each consumer filling his own little balloons with CO_2 to be processed by his municipality. The program has to be tackled upstream.[68]

Although he's right, I wonder if we would have a deeper appreciation of the problem if each of us did have to take care of our carbon dioxide wastes, or at least directly pay for them. In any case, by "tackled upstream," Marchetti meant scrubbing carbon dioxide out of flue gas emitted by power plants, waste incinerators, and industry. It can be done now with solvents such as monoethanolamine, solid absorbents, or separation by membranes.[69] The microbial way is to use the photosynthetic activity of microalgae to take up carbon dioxide from flue gas bubbled into water. There are several problems with algal scrubbers, however, starting with the harsh environment facing algae trying to grow in flue-gas-infused water. Flue gas is hot, reaching 100 degrees Celsius, with high amounts of carbon dioxide (5 to 15 percent) and enough nitrogen oxides and sulfur oxides to turn water into a concentrated acid where few microbes can survive. Research has focused on finding algae that can grow well enough under those harsh conditions to be effective carbon dioxide scrubbers. Although many studies concentrate on single algal species, investigators in Australia took a different approach.[70] They added natural microbial communities from a lake and creek to 30-liter conical transparent polyethylene bags and then over time exposed the microbes to increasing levels of flue gas from a small coal power plant. After several months, an algal community dominated by *Desmodesmus* became acclimated to 100 percent flue gas, although the water had to be buffered to reduce its acidity. Among several reasons why power plant operators have not rushed to install algal-carbon dioxide scrubbers is the area required by them to be effective. The Australian investigators calculated that to remove 90 percent of the gas emitted by an average-size coal power plant, an algae-filled pond would have to cover 224 square kilometers, equal to an area of about 15 by 15 kilometers.

A few carbon capture and storage (CCS) facilities have been built, and more are on the way. According to the 2022 report from the Global CCS Institute, 30 facilities already exist around the world, another 11 are being constructed, and over 150 are in development.[71] When all existing and planned CCS facilities are in operation, they would remove over 240 million tons of carbon dioxide each year. That seems like a big number, but it's nothing when stacked against the over 37,000 million tons of carbon dioxide now released annually by human activity.[72] Smaller CCS devices closer to Marchetti's balloons could help. In 2021 a carbon scrubber was installed to capture the carbon dioxide released from the heating system of an apartment building in the Upper West Side of Manhattan.[73] The scrubber doesn't use algae nor were microbes mentioned in the CCS report. Some of the many CCS facilities that will be needed in the coming decades could use algae either in open ponds or in closed bioreactor towers. Companies are now making algal bioreactors to order. According to the website of one based in Denmark, ALGIECEL makes a "plug & play photobioreactor" that "fits into standard shipping containers," suitable for removing the carbon dioxide released by a small industry.[74] A cynic would say these small bioreactors do more for a company's public image ("greenwashing" comes to mind) than for the environment, but they shouldn't be dismissed just because of their size. Enough of them working efficiently could make a difference.

Other negative emission technologies operate on a much larger scale than a single home or even a big power plant. Some aim to remove atmospheric carbon dioxide from the entire planet. A couple of these negative emission schemes are quite controversial for one or more reasons. They may work in theory yet are impossible to implement in today's world with its political and socioeconomic barriers, or even if they could be implemented, there may be unintended consequences. Probably more so than the other solutions introduced so far in this chapter, the negative emission technologies to be discussed next have been criticized as a diversion from the real solution to the climate change problem: stop using fossil fuels. Yet, as pointed out before, we have no choice but to remove atmospheric carbon dioxide if global warming is to be minimized. Rather than trying to judge the worthiness of the following negative emission schemes, my goal here is to provide the microbiological background so that readers are better informed to reach their own conclusion.

Geoengineering the Land

Marchetti was the first to use the term "geoengineering" as it's now employed in the climate change literature,[75] although it appears only in the title of his 1977 paper. After discussing removing carbon dioxide from power plant flue gas, Marchetti went on to talk about storing the recovered carbon in the deep ocean. Today geoengineering is much bigger and broader than that. It's defined as the deliberate, large-scale manipulation of soils, oceans, or the atmosphere to combat climate change. Some of these manipulations, such as placing giant umbrella-like structures in outer space to shade Earth from the sun, seem to have been taken straight from science fiction. Kenneth Roy recently argued that a solar shield situated at the Sun-Earth L1 point, about 1.5 million kilometers from Earth, is safer than other geoengineering solutions.[76] It does seem less intrusive than other ways proposed to block sunlight, such as shooting sulfur dioxide, soot, or aluminum particles into the stratosphere to create sunlight-blocking aerosols.[77] Climate scientists know stratospheric aerosols would be effective after seeing how the world cooled following the injection of several thousand tons of sulfur dioxide and ash by the eruption of Mount Pinatubo in 1991. A sunscreen from microbes is dimethyl sulfide (Chapter 5). Geoengineering with the sulfur gas would entail fertilizing the Southern Ocean with iron in order to enhance production of dimethyl sulfide and formation of aerosols and clouds. But these geoengineering solutions can't be the complete answer, even assuming they would work. Solar shields, stratospheric aerosols, or thicker cloud cover wouldn't help with ocean acidification (Chapter 6) and other problems caused by rising atmospheric carbon dioxide.

Other geoengineering solutions are negative emission technologies writ large (Fig. 9.4). One is to have more of the best carbon dioxide scrubber on the planet, better than anything humans could invent: trees. Trees are a big

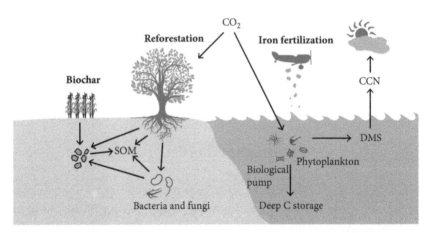

Figure 9.4 Some geoengineering solutions to draw down atmospheric carbon dioxide (CO_2) or to stimulate the release of dimethyl sulfide (DMS) and the creation of cloud condensation nuclei (CCN). Biochar promotes formation of soil organic matter (SOM) by slowing the degradation of organic chemicals released by microbes and terrestrial plants. Reforestation stores carbon in soils and contributes organic chemicals used by microbes to produce SOM. Iron fertilization stimulates phytoplankton growth and strengthens the biological pump, which exports carbon to deep waters and sediments out of contact with the atmosphere.

reason why terrestrial habitats already soak up about 29 percent of carbon dioxide released by human activity (Chapter 3). Governments, nonprofit organizations, and companies are now planting trees on land once covered with forests (reforestation) or not (afforestation) at a scale implied by the name of some of these efforts. The Nature Conservancy, for example, has its Plant a Billion Trees project.[78] Recent studies have questioned whether reforestation will store as much carbon as originally advertised,[79] and it doesn't make much sense to plant trees where natural grasslands once flourished. In those regions, grasses would store carbon more sustainably. Yet reforestation would help, if enough is done to ensure the trees' survival and growth. That depends in part on microbes. Nearly all trees and other land vegetation rely on mycorrhizal fungi living symbiotically with roots to supply the nutrients and water essential for plants to flourish (Chapter 3). If you pay a company to plant some trees on your behalf to offset the carbon released during a transAtlantic flight, you should ask the company to ensure the trees get the right fungi (commercial mixes are available). With the proper mycorrhizal fungi, trees can grow three times faster[80] and potentially store that much more carbon.

Microbes help trees store carbon in another way, by forming soil organic matter (SOM). The carbon in SOM comes from the carbon dioxide used in photosynthesis to make the organic chemicals that plants need for growth and reproduction. Once in soil, the plant organics are then transmogrified by bacteria and fungi into SOM organic chemicals that resist further degradation for thousands of years (Chapter 3). A study published in 2022 looked into how much carbon dioxide could be stored by reforestation and better forest management and concluded that SOM would account for about 20 percent of the stored carbon with the remaining 80 percent in plant biomass.[81] Much of that plant carbon would also likely be turned eventually into SOM. Whether the organic carbon is stored in SOM or in plant debris, it was made by trees with substantial help from fungi and bacteria.

Another intervention could create more SOM and take more carbon dioxide out of the atmosphere. Along with adding biochar to manure or rice fields to lower methane emissions, folding biochar into soils is thought to sequester carbon as well as to absorb pollutants and to increase crop yields by improving soil quality. The term "biochar" was first used in 1999 to describe the activated carbon prepared from burnt sorghum grain,[82] yet its use in agriculture has been around for years. Since the first millennium BCE, indigenous people of Amazonia have added charcoal along with plant debris to soils, forming terra preta, in order to improve soil fertility.[83] Being much harder to degrade than unburnt wood or other plant material, biochar itself is a form of carbon storage. It sequesters even more carbon by helping to make SOM.[84] The organic chemicals from plants, bacteria, and fungi absorb onto biochar, transforming them into SOM components that resist microbial degradation for decades to millennia. Widespread application of biochar could supply as much as 35 percent of the negative emissions needed by IPCC scenarios to limit global warming to under 2 degrees Celsius.[85]

Assessing the benefits of another large store of organic matter, peat, is more complicated. As mentioned in Chapter 3, peat is organic matter from wetland plants that has built up over thousands of years in waterlogged soil where the plant matter has been protected from microbial degradation by acidic conditions and low oxygen levels. As a result, peat holds a lot of carbon, about a third of what is in soils, much more than what is in trees and other vegetation. Draining peatlands to recover the land for agriculture or forestry can release carbon dioxide when the peat organic carbon becomes available for degradation as oxygen levels increase, but it can also reduce emissions of two other, powerful greenhouse gases, methane and nitrous oxide, produced in waterlogged soil by anaerobic archaea and bacteria. So would it be best to continue to drain peatlands? No. It turns out that the higher release of carbon dioxide by drained peatlands exceeds the benefit of lower methane and nitrous oxide emissions.[86] A thorough cost-benefit analysis is complicated, involving factors like the height of the water table, which can drop as climate change alters the hydrological cycle. But in the end, rather than getting rid of them, more should be done to restore peatlands and to ensure they continue to sequester carbon.

Along the same line, we should keep other types of wetlands and restore at least some of those that have been drained. Any benefit from lowering methane and nitrous oxide emissions by draining wetlands is outweighed by the loss of a powerful way to sequester carbon.[87] One study estimated that restoring wetlands by rewetting them could reduce greenhouse gas emissions equal to about 10 percent of what humans produce. There are other reasons to preserve wetlands. They serve as habitat for wildlife and as biofilters that remove nutrients and other chemicals which otherwise could pollute rivers, lakes, and drinking water. Although now being done to promote carbon storage, restoration of a peatland in Scotland was started to increase habitat suitable for birds.[88] Whether as wildlife habitats, biofilters, or carbon stores, wetlands are too valuable to lose and are often worth the cost of restoring them.

Restoring peatlands and other wetlands, reforestation, and biochar applications all aim to increase the amount of carbon sequestered on land. More low-key than planting a billion trees, just using good soil management practices could increase soil quality and the amount of carbon sequestered in soils.[89] The right soil management could lead to the formation of more necromass, the dead remains of microbes that resist degradation, making it an effective way to store carbon in soils and to enhance soil fertility. The right amount of grazing on grasslands could increase SOM, although even more carbon could be sequestered if grasslands are restored to their natural state.[90] All these strategies fit with the 4 per 1000 initiative, whose goal is to increase soil carbon by 4 parts per thousand or 0.4 percent.[91]

However, soils cover less than 10 percent of Earth's surface and are unlikely to be the only answer. More storage space for carbon can be found in the other place in the world where carbon stocks are already high: the ocean.

Geoengineering the Ocean

Oceans store even more carbon than soils do and absorb nearly as much anthropogenic carbon dioxide. Perhaps the oceans could do even more after some geoengineering. Some of the ocean engineering ideas are as wild as putting sun shields in outer space. One idea is to use wave-driven pumps to increase upwelling or its reverse, downwelling.[92] More upwelling of deep waters would bring nutrients to the nutrient-poor, sunlit surface, stimulating more vigorous phytoplankton growth and a stronger biological pump that sends carbon to deep waters out of contact with the atmosphere (Chapter 4). During downwelling, surface water and all its carbon pools and heat are pushed down to deep water naturally by winds or in a geoengineering project, by pumps. Another ocean geoengineering scheme, large-scale cultivation of macroalgae such as kelp, is like reforestation on land. Cultivated macroalgae, a multibillion-dollar industry, is now used for fish and human food, fertilizer, and cosmetics, but if used to store carbon, the macroalgae would have to be buried somehow to prevent any carbon dioxide escaping back into the atmosphere. None of these ocean geoengineering ideas have been tested adequately to see if they would work.

The one that has been studied extensively is controversial. The idea is to enhance phytoplankton growth and to strengthen the biological pump by fertilizing the ocean. Most climate scientists have focused on fertilizing with iron those oceanic waters, known as high nutrient-low chlorophyll (HNLC) regions, where low iron concentrations prevent vigorous phytoplankton growth, resulting in low chlorophyll, despite high concentrations of nutrients. Of the few HNLC waters around the world, the biggest is the Southern Ocean. Some evidence that iron fertilization would work comes from studies arguing that ice ages over the last 650,000 years were caused in part by iron inputs into the Southern Ocean that stimulated phytoplankton enough to strengthen the biological pump and draw down atmospheric carbon dioxide (Chapter 4). John Martin, the oceanographer who came up with the iron hypothesis, famously quipped, "Give me half a tanker of iron and I'll give you the next ice age." Fertilization would require only small amounts of iron because phytoplankton need only a bit of the metal, unlike other nutrients. For geoengineering purposes, iron fertilization of the ocean would consume less than a percent of the current world supply, whereas fertilization with nitrogen or phosphorus would use up 30 to 40 percent.[93]

Oceanographers and climate scientists know a lot about ocean iron fertilization because it's been done several times already. Thirteen scientific studies conducted between 1993 and 2009 released 350 to 4000 kilograms of iron sulfate off the stern of ships while they sailed over 25 to 300 square kilometers of the Southern Ocean and other HNLC waters. In terms of science, the studies were spectacularly successful. The added iron stimulated primary production by as much as 10-fold, leading to massive phytoplankton blooms. The studies clearly showed how iron sets phytoplankton growth in HNLC waters, but they were less conclusive about the effectiveness of iron fertilization as a geoengineering solution. Because everything about oceanographic research, from the personnel to the ships, is so expensive, the 13 iron fertilization studies couldn't follow the phytoplankton bloom long enough to explore carbon export and to demonstrate conclusively that iron fertilization enhances carbon sequestration in the deep ocean. Still, iron fertilization looks like it would work in removing at least some carbon dioxide from the atmosphere.

By the 2000s, iron fertilization even for science attracted the ire of environmentalists, yet it was a small project without much science that was more controversial. In 2012, 100 tons of iron sulfate were dumped by Planktos, a company based in San Francisco, California, into coastal waters off Old Massett, a fishing village of 1000 people, in British Columbia.[94] Rather than for carbon storage, the villagers wanted to revive their salmon fishery, thinking correctly that more phytoplankton means more fish. The iron addition did stimulate a phytoplankton bloom, yet it didn't seem to do much for the salmon, although large harvests of sockeye salmon have been connected to the iron emitted by the eruption of the Kasatochi volcano in 2008.[95] The Planktos project was controversial because it was done without prior approval of the Canadian federal government.[96] Legal questions and concerns about governmental oversight and verifying the results are some of the arguments against ocean fertilization.

The costs and benefits of ocean fertilization were closely examined in a 2022 report from the U.S. National Academies of Sciences, Engineering, and Medicine,[97] which I've relied on heavily here because it's an up-to-date, balanced review of many previous studies published in the peer-reviewed literature. The report thoroughly studied the impacts of iron fertilization on the marine ecosystem and presented evidence, I think it's fair to say, that the side effects may not be as bad as some people fear, although the benefits may not be as big as others hope. To be clear, the report strives to be neutral and doesn't advocate fertilization or any other oceanic geoengineering scheme. Ocean fertilization should be seriously considered, but I think other strategies should be tried first or implemented more broadly than they currently are.

Power of the Unseen

Ocean fertilization is a dramatic example of the potential power of harnessing microbes to help solve the greenhouse gas problem. This chapter has presented several other, less extreme solutions that exemplify microbial potency. More successful than any medieval alchemist, microbes can turn the dross of manure piles, landfills, and wastewater into valuable biomethane and electricity. Trees are the obvious stars of reforestation efforts, yet each tree in The Nature Conservancy's Plant a Billion Trees project depends on nutrients supplied by symbiotic fungi and bacteria. For biochar applications or the 4 per 1000 initiative to succeed, microbes must produce even more refractory SOM that resists degradation and stores carbon for centuries. There is a limit, however, to what microbes can do alone. With all these potential solutions, their success depends on us taking full advantage of the power of microbes. Just as reducing fossil fuels—the best powerful solution to the greenhouse gas problem—depends on us, so too do the microbial solutions.

Although it's only escapist entertainment, *Mad Max Beyond Thunderdome* does bring to light a bit of truth. We see the pigs in the subterranean refinery, but we can't see what really fuels Bartertown: the microbes, the archaea that actually produce the biomethane. So it is for the real world. From redwoods and dandelions to elephants and whales, all megaflora and megafauna exist thanks to microbes. That's true also for us, most intimately through the bacteria and other microbes in the human microbiome. All of the biosphere has been supported by microbes for billions of years, ever since life began on Earth. Just as the biosphere can only be understood by looking at microbes, so too for climate change. Once only specialists working in obscure corners of microbial ecology and biogeochemistry appreciated the crucial role of microbes in the biosphere, but now anyone who wants to understand climate change needs to know about microbes. More urgently than ever, understanding and solving the most serious environmental problem facing us today depends on the smallest organisms: the microbes.

Acknowledgments

Several people provided invaluable input and comments at various stages of writing this book. I thank the following people who reviewed early drafts of one or more chapters: Grant Allen, David Archer, Chandranath Basak, David Bastviken, Tim Bertram, Ben Bond-Lamberty, Klaus Butterbach-Bahl, Doug Capone, Jorge Cortes, Jeana Drake, Hugh Ducklow, Eliza Harris, Dave Hutchins, Jinshi Jian, John Kessler, Joan Kleypas, Naomi M. Levine, Bruce Logan, Ashish Malik, Andrew McGuire, Trish Quinn, Evelyn Sherr, Josh Schimel, Eric Slessarev, Bess Ward, and Jamie Wilson. Thanks are also due to the following people who provided information or helped in other ways: Humberto Blanco, Colleen Cavanaugh, Ken Johnson, Chaoqun Lu, Anoop Mahajan, Matt Oliver, Carol Robinson, Johannes Rousk, Carolyn Ruppel (who suffered through a video call and many emails from me), Rodrigo Vargas, and Dana Veron. Lisa Stern deserves special thanks for her input about the overall direction of the book and for reviewing a chapter. Finally, Jean Brodeur and Tiffany Straza read every word of the book and pushed me to do better.

This book would not have been possible without the papers, books, and other aid provided by the Morris Library at the University of Delaware and, through its interlibrary loan program, by several other libraries around the world. Libraries are the underappreciated foundation of careful scholarship.

Notes

Chapter 1

1. Malavika Vyawahare, "Hawaii First to Harness Deep-Ocean Temperatures for Power," *Scientific American* (August 27, 2015), https://www.scientificamerican.com/article/hawaii-first-to-harness-deep-ocean-temperatures-for-power/, accessed March 31, 2023.
2. Charles D. Keeling, "Rewards and Penalties of Monitoring the Earth," *Annual Review of Energy and the Environment* 23 (1998): 25–82. doi: 10.1146/annurev.energy.23.1.25.
3. https://gml.noaa.gov/ccgg/about.html, accessed November 17, 2021.
4. Charles D. Keeling, "The Concentration and Isotopic Abundances of Carbon Dioxide in the Atmosphere," *Tellus* 12 (1960): 200–03. doi: 10.1111/j.2153-3490.1960.tb01300.x.
5. Paul G. Falkowski, *Life's Engines: How Microbes Made Earth Habitable*, Science Essentials (Princeton, NJ: Princeton University Press, 2015).
6. Nick Lane, "The Unseen World: Reflections on Leeuwenhoek (1677) 'Concerning Little Animals,'" *Philosophical Transactions of the Royal Society B: Biological Sciences* 370 (2015): . doi: 10.1098/rstb.2014.0344.
7. C. R. Woese and G. E. Fox, "Phylogenetic Structure of Prokaryotic Domain—Primary Kingdoms," *Proceedings of the National Academy of Sciences* 74 (1977): 5088–90.
8. Thiago Rodrigues-Oliveira, Florian Wollweber, Rafael I. Ponce-Toledo, Jingwei Xu, Simon K. M. R. Rittmann, Andreas Klingl, Martin Pilhofer, et al., "Actin Cytoskeleton and Complex Cell Architecture in an Asgard Archaeon," *Nature* 613 (2022): 332–39. doi: 10.1038/s41586-022-05550-y.
9. Yinon M. Bar-On, Rob Phillips, and Ron Milo, "The Biomass Distribution on Earth," *Proceedings of the National Academy of Sciences* 115 (2018): 6506–11. doi: 10.1073/pnas.1711842115.
10. Patricia Sánchez-Baracaldo, Giorgio Bianchini, Jamie D. Wilson, and Andrew H. Knoll, "Cyanobacteria and Biogeochemical Cycles through Earth History," *Trends in Microbiology* 30 (2022): 143–57. doi: 10.1016/j.tim.2021.05.008.
11. A. Scott Denning, "Where Has All the Carbon Gone?" *Annual Review of Earth and Planetary Sciences* 50 (2022): 55–78. doi: 10.1146/annurev-earth-032320-092010. Quote on p. 57.
12. P. Friedlingstein, M. O'Sullivan, M. W. Jones, R. M. Andrew, L. Gregor, J. Hauck, C. Le Quéré, et al. "Global Carbon Budget 2022," *Earth System Science Data* 14 (2022): 4811–900. doi: 10.5194/essd-14-4811-2022.
13. https://gml.noaa.gov/ccgg/trends/, accessed March 9, 2023.
14. M. Etminan, G. Myhre, E. J. Highwood, and K. P. Shine, "Radiative Forcing of Carbon Dioxide, Methane, and Nitrous Oxide: A Significant Revision of the Methane

Radiative Forcing," *Geophysical Research Letters* 43 (2016): 12614–623. doi: 10.1002/2016GL071930.

15. Toby Tyrrell, *On Gaia: A Critical Investigation of the Relationship between Life and Earth* (Princeton, NJ: Princeton University Press, 2013), p. 3.

16. Leah Aronowsky, "Gas Guzzling Gaia, Or: A Prehistory of Climate Change Denialism," *Critical Inquiry* 47 (2021): 306–27. doi: 10.1086/712129.

17. James Rodger Fleming, *Historical Perspectives on Climate Change* (New York: Oxford University Press, 1998).

18. Gilbert N. Plass, "The Carbon Dioxide Theory of Climatic Change," *Tellus* 8 (1956): 140–54. doi: 10.1111/j.2153-3490.1956.tb01206.x.

19. P. Forster, T. Storelvmo, K. Armour, W. Collins, J.-L. Dufresne, D. Frame, D. J. Lunt, et al., "The Earth's Energy Budget, Climate Feedbacks, and Climate Sensitivity," in *Climate Change 2021: The Physical Science Basis. Contribution of Working Group I to the Sixth Assessment Report of the Intergovernmental Panel on Climate Change*, ed. V. Masson-Delmotte, P. Zhai, A. Pirani, S. L. Connors, C. Péan , S. Berger, N. Caud, Y. Chen, L. Goldfarb, M. I. Gomis, M. Huang, K. Leitzell, E. Lonnoy, J. B. R. Matthews, T. K. Maycock, T. Waterfield, O. Yelekçi, R. Yu, and B. Zhou (Cambridge: Cambridge University Press, 2021), pp. 923–1054.

20. D. Chen, M. Rojas, B. H. Samset, K. Cobb, A. Diongue Niang, P. Edwards, S. Emori, et al., "Framing, Context, and Methods," in *Climate Change 2021: The Physical Science Basis. Contribution of Working Group I to the Sixth Assessment Report of the Intergovernmental Panel on Climate Change*, ed. V. Masson-Delmotte, P. Zhai, A. Pirani, S.L. Connors, C. Péan , S. Berger, N. Caud, Y. Chen, L. Goldfarb, M. I. Gomis, M. Huang, K. Leitzell, E. Lonnoy, J. B. R. Matthews, T. K. Maycock, T. Waterfield, O. Yelekçi, R. Yu, and B. Zhou (Cambridge: Cambridge University Press, 2021), pp. 147–286.

21. Blake D. Key, Robert D. Howell, and Craig S. Criddle, "Fluorinated Organics in the Biosphere," *Environmental Science & Technology* 31 (1997): 2445–54. doi: 10.1021/es961007c.

Chapter 2

1. The presentation is mentioned in M. A. Matthews, "The Earth's Carbon Cycle," *New Scientist* 6 (1959): 644–46. A recording of another presentation by Teller is available at https://kuci.org/wp/podcast/edward-teller-speech/, accessed December 15, 2021. Teller discussed the impact of fossil fuel consumption and global warming in Allan Nevins, Robert G. Dunlop, Edward Teller, Edward S. Mason, and Herbert Hoover, Jr., *Energy and Man: A Symposium* (New York: Appleton-Century-Crofts, 1960).

2. S. K. Gulev, P. W. Thorne, J. Ahn, F. J. Dentener, C. M. Domingues, S. Gerland, D. Gong, et al., "Changing State of the Climate System," in *Climate Change 2021: The Physical Science Basis. Contribution of Working Group I to the Sixth Assessment Report of the Intergovernmental Panel on Climate Change*, ed. V. Masson-Delmotte, P. Zhai,

A. Pirani, S. L. Connors, C. Péan, S. Berger, N. Caud, Y. Chen, L. Goldfarb, M. I. Gomis, M. Huang, K. Leitzell, E. Lonnoy, J. B. R. Matthews, T. K. Maycock, T. Waterfield, O. Yelekçi, R. Yu, and B. Zhou (Cambridge: Cambridge University Press, 2021), pp. 287–422.

3. Merritt R. Turetsky, Brian Benscoter, Susan Page, Guillermo Rein, Guido R. van der Werf, and Adam Watts, "Global Vulnerability of Peatlands to Fire and Carbon Loss," *Nature Geoscience* 8 (2015): 11–14. doi: 10.1038/ngeo2325.

4. Fred T. Mackenzie and Abraham Lerman, *Carbon in the Geobiosphere: Earth's Outer Shell* (Dordrecht: Springer, 2006), https://doi.org/10.1007/1-4020-4238-8.

5. William B. Whitman, David C. Coleman, and William J. Wiebe, "Prokaryotes: The Unseen Majority," *Proceedings of the National Academy of Sciences* 95 (1998): 6578–83. doi: 10.1073/pnas.95.12.6578.

6. Yinon M. Bar-On, Rob Phillips, and Ron Milo, "The Biomass Distribution on Earth," *Proceedings of the National Academy of Sciences* 115 (2018): 6506–11. doi: 10.1073/pnas.1711842115.

7. Henrik Lundegardh, "Carbon Dioxide Evolution of Soil and Crop Growth," *Soil Science* 23 (1927): 417–53.

8. Jukka Pumpanen, Pasi Kolari, Hannu Ilvesniemi, Kari Minkkinen, Timo Vesala, Sini Niinistö, Annalea Lohila, et al., "Comparison of Different Chamber Techniques for Measuring Soil CO_2 Efflux," *Agricultural and Forest Meteorology* 123 (2004): 159–76. https://doi.org/10.1016/j.agrformet.2003.12.001.

9. J. Jian, R. Vargas, K. Anderson-Teixeira, E. Stell, V. Herrmann, M. Horn, N. Kholod, et al., "A Restructured and Updated Global Soil Respiration Database (SRDB-V5)," *Earth System Science Data* 13 (2021): 255–67. doi: 10.5194/essd-13-255-2021.

10. Jinshi Jian, Vanessa Bailey, Kalyn Dorheim, Alexandra G. Konings, Dalei Hao, Alexey N. Shiklomanov, Abigail Snyder, et al., "Historically Inconsistent Productivity and Respiration Fluxes in the Global Terrestrial Carbon Cycle," *Nature Communications* 13 (2022): 1733. doi: 10.1038/s41467-022-29391-5.

11. Jinshi Jian, Max Frissell, Dalei Hao, Xiaolu Tang, Erin Berryman, and Ben Bond-Lamberty, "The Global Contribution of Roots to Total Soil Respiration," *Global Ecology and Biogeography* 31 (2022): 685–99. https://doi.org/10.1111/geb.13454.

12. Mengguang Han, Jiguang Feng, Ying Chen, Lijuan Sun, Liangchen Fu, and Biao Zhu, "Mycorrhizal Mycelial Respiration: A Substantial Component of Soil Respired CO_2," *Soil Biology and Biochemistry* 163 (2021): 108454. https://doi.org/10.1016/j.soilbio.2021.108454.

13. X. Tang, S. Fan, M. Du, W. Zhang, S. Gao, S. Liu, G. Chen, et al., "Spatial and Temporal Patterns of Global Soil Heterotrophic Respiration in Terrestrial Ecosystems," *Earth System Science Data* 12 (2020): 1037–51. doi: 10.5194/essd-12-1037-2020.

14. Jian et al., "Inconsistent Productivity."

15. Carol Robinson, "Microbial Respiration, the Engine of Ocean Deoxygenation," *Frontiers in Marine Science* 5 (2019). doi: 10.3389/fmars.2018.00533.

16. C. Robinson and Peter J. le B. Williams, "Respiration and Its Measurement in Surface Marine Waters," in *Respiration in Aquatic Ecosystems*, ed. Paul A. del Giorgio and Peter J. le B. Williams (Oxford: Oxford University Press, 2005), pp. 147–80.

17. Eric L. Mills, *Biological Oceanography: An Early History, 1870–1960* (Toronto: University of Toronto Press, 2012), https://www.jstor.org/stable/10.3138/978144 2663053.
18. Paul A. del Giorgio and Carlos M. Duarte, "Respiration in the Open Ocean," *Nature* 420 (2002): 379–84. doi: 10.1038/nature01165.
19. Robinson and Williams, "Respiration."
20. https://biogeochemical-argo.org/, accessed December 23, 2021. See also Hervé Claustre, Kenneth S. Johnson, and Yuichiro Takeshita, "Observing the Global Ocean with Biogeochemical-Argo," *Annual Review of Marine Science* 12 (2020): 23–48. doi: 10.1146/annurev-marine-010419-010956.
21. https://argo.ucsd.edu/, accessed December 22, 2021.
22. Kenneth S. Johnson and Mariana B. Bif, "Constraint on Net Primary Productivity of the Global Ocean by Argo Oxygen Measurements," *Nature Geoscience* 14 (2021): 769–74. doi: 10.1038/s41561-021-00807-z.
23. https://www.mbari.org/about/history/, accessed January 12, 2022.
24. Allison R. Moreno, Catherine A. Garcia, Alyse A. Larkin, Jenna A. Lee, Wei-Lei Wang, J. Keith Moore, Francois W. Primeau, et al., "Latitudinal Gradient in the Respiration Quotient and the Implications for Ocean Oxygen Availability," *Proceedings of the National Academy of Sciences* 117 (2020): 22866–72. doi: 10.1073/pnas.2004986117.
25. Fernanda Henderikx Freitas, Angelicque E. White, and Paul D. Quay, "Diel Measurements of Oxygen- and Carbon-Based Ocean Metabolism across a Trophic Gradient in the North Pacific," *Global Biogeochemical Cycles* 34 (2020): e2019GB006518. https://doi.org/10.1029/2019GB006518.
26. P. Friedlingstein, M. O'Sullivan, M. W. Jones, R. M. Andrew, L. Gregor, J. Hauck, C. Le Quéré, et al. "Global Carbon Budget 2022." *Earth System Science Data* 14 (2022): 4811–900. doi: 10.5194/essd-14-4811-2022.
27. https://fluxnet.org/, accessed December 22, 2021. See also Dennis D. Baldocchi, "How Eddy Covariance Flux Measurements Have Contributed to Our Understanding of Global Change Biology," *Global Change Biology* 26 (2020): 242–60. https://doi.org/10.1111/gcb.14807.
28. Grayson Badgley, Leander D. L. Anderegg, Joseph A. Berry, and Christopher B. Field, "Terrestrial Gross Primary Production: Using NIRV to Scale from Site to Globe," *Global Change Biology* 25 (2019): 3731–40. doi: 10.1111/gcb.14729.
29. Jian et al., "Inconsistent Productivity."
30. Sallie W. Chisholm, Robert J. Olson, Erik R. Zettler, Ralf Goericke, John B. Waterbury, and Nicholas A. Welschmeyer, "A Novel Free-Living Prochlorophyte Abundant in the Oceanic Euphotic Zone," *Nature* 334 (1988): 340–43. doi: 10.1038/334340a0.
31. https://www.learner.org/series/the-habitable-planet-a-systems-approach-to-enviro nmental-science/oceans/interview-with-penny-chisholm/, accessed March 31, 2023.
32. F. Partensky, W. R. Hess, and D. Vaulot, "*Prochlorococcus*, a Marine Photosynthetic Prokaryote of Global Significance," *Microbiology and Molecular Biology Reviews* 63 (1999): 106–27. doi:10.1128/MMBR.63.1.106-127.1999.
33. Steven J. Biller, Paul M. Berube, Debbie Lindell, and Sallie W. Chisholm, "*Prochlorococcus*: The Structure and Function of Collective Diversity," *Nature Reviews Microbiology* 13 (2015): 13–27. doi: 10.1038/nrmicro3378.

34. Peng Zhou, Li Wang, Hai Liu, Chunyan Li, Zhimin Li, Jinxiang Wang, and Xiaoming Tan, "CyanoOmicsDB: An Integrated Omics Database for Functional Genomic Analysis of Cyanobacteria," *Nucleic Acids Research* 50 (2021): D758–D64. doi: 10.1093/nar/gkab891.

35. Gerald A. Tuskan, Andrew T. Groover, Jeremy Schmutz, Stephen Paul DiFazio, Alexander Myburg, Dario Grattapaglia, Lawrence B. Smart, et al., "Hardwood Tree Genomics: Unlocking Woody Plant Biology," *Frontiers in Plant Science* 9 (2018). doi: 10.3389/fpls.2018.01799.

36. Yibin Huang, David Nicholson, Bangqin Huang, and Nicolas Cassar, "Global Estimates of Marine Gross Primary Production Based on Machine Learning Upscaling of Field Observations," *Global Biogeochemical Cycles* 35 (2021): e2020GB006718. https://doi.org/10.1029/2020GB006718.

37. Howard R. Gordon, "Some Reflections on Thirty-Five Years of Ocean Color Remote Sensing," in *Oceanography from Space, Revisited*, ed. Vittorio Barale, J. F. R. Gower, and L. Alberotanza (Dordrecht: Springer, 2010), pp. 289–306.

38. Gordon, "Some Reflections," p. 295.

39. Carlos M. Duarte, Aurore Regaudie-de-Gioux, Jesús M. Arrieta, Antonio Delgado-Huertas, and Susana Agustí, "The Oligotrophic Ocean Is Heterotrophic," *Annual Review of Marine Science* 5 (2013): 551–69. doi: 10.1146/annurev-marine-121211-172337.

40. Peter J. le B. Williams, Paul D. Quay, Toby K. Westberry, and Michael J. Behrenfeld, "The Oligotrophic Ocean Is Autotrophic," *Annual Review of Marine Science* 5 (2013): 535–49. doi: 10.1146/annurev-marine-121211-172335.

41. C. B. Field, M. J. Behrenfeld, J. T. Randerson, and P. Falkowski, "Primary Production of the Biosphere: Integrating Terrestrial and Oceanic Components," *Science* 281 (1998): 237–40. doi: 10.1126/science.281.5374.237.

42. Bar-On et al., "Biomass on Earth."

43. Frédérik Saltré and Corey J. A. Bradshaw, "Climate Explained: What Was the Medieval Warm Period?," *The Conversation* (April 20, 2021), https://theconversation.com/climate-explained-what-was-the-medieval-warm-period-155294, accessed Feburary 9, 2022.

Chapter 3

1. P. Friedlingstein, M. O'Sullivan, M. W. Jones, R. M. Andrew, L. Gregor, J. Hauck, C. Le Quéré, et al., "Global Carbon Budget 2022," *Earth System Science Data* 14 (2022): 4811–900. doi: 10.5194/essd-14-4811-2022.

2. Pierre-Marc Delaux and Sebastian Schornack, "Plant Evolution Driven by Interactions with Symbiotic and Pathogenic Microbes," *Science* 371 (2021): eaba6605. doi:10.1126/science.aba6605.

3. Miguel A. Naranjo-Ortiz and Toni Gabaldón, "Fungal Evolution: Major Ecological Adaptations and Evolutionary Transitions," *Biological Reviews* 94 (2019): 1443–76. https://doi.org/10.1111/brv.12510.

4. Noah W. Sokol, Eric Slessarev, Gianna L. Marschmann, Alexa Nicolas, Steven J. Blazewicz, Eoin L. Brodie, Mary K. Firestone, et al., "Life and Death in the Soil Microbiome: How Ecological Processes Influence Biogeochemistry," *Nature Reviews Microbiology* 20 (2022): 415–30. doi: 10.1038/s41579-022-00695-z.

5. Jing-Ke Weng and Clint Chapple, "The Origin and Evolution of Lignin Biosynthesis," *New Phytologist* 187 (2010): 273–85. https://doi.org/10.1111/j.1469-8137.2010.03327.x.

6. Hongwei Wang and Edgar Chambers IV, "Sensory Characteristics of Various Concentrations of Phenolic Compounds Potentially Associated with Smoked Aroma in Foods," *Molecules* 23 (2018): 780. doi: 10.3390/molecules23040780.

7. Łukasz Pawlik, Brian Buma, Pavel Šamonil, Jiří Kvaček, Anna Gałązka, Petr Kohout, and Ireneusz Malik, "Impact of Trees and Forests on the Devonian Landscape and Weathering Processes with Implications to the Global Earth's System Properties—A Critical Review," *Earth-Science Reviews* 205 (2020): 103200. https://doi.org/10.1016/j.earscirev.2020.103200.

8. William E. Stein, Christopher M. Berry, Linda VanAller Hernick, and Frank Mannolini, "Surprisingly Complex Community Discovered in the Mid-Devonian Fossil Forest at Gilboa," *Nature* 483 (2012): 78–81. doi: 10.1038/nature10819.

9. W. A. Dimichele, "Carboniferous Coal-Swamp Forests," in *Palaeobiology II*, ed. D. E. G. Briggs and Peter R. Crowther (2001) (Hoboken, NJ: Wiley-Blackwell), pp. 79–82.

10. Russell J. Garwood and Gregory D. Edgecombe, "Early Terrestrial Animals, Evolution, and Uncertainty," *Evolution: Education and Outreach* 4 (2011): 489–501. doi: 10.1007/s12052-011-0357-y.

11. Matthew P. Nelsen, William A. DiMichele, Shanan E. Peters, and C. Kevin Boyce, "Delayed Fungal Evolution Did Not Cause the Paleozoic Peak in Coal Production," *Proceedings of the National Academy of Sciences* 113 (2016): 2442–47. doi: 10.1073/pnas.1517943113.

12. Dimitrios Floudas, Manfred Binder, Robert Riley, Kerrie Barry, Robert A. Blanchette, Bernard Henrissat, Angel T. Martínez, et al., "The Paleozoic Origin of Enzymatic Lignin Decomposition Reconstructed from 31 Fungal Genomes," *Science* 336 (2012): 1715–19. doi: 10.1126/science.1221748.

13. Isabel Patricia Montañez, "A Late Paleozoic Climate Window of Opportunity," *Proceedings of the National Academy of Sciences* 113 (2016): 2334–36. doi: 10.1073/pnas.1600236113.

14. Grzegorz Janusz, Anna Pawlik, Justyna Sulej, Urszula Świderska-Burek, Anna Jarosz-Wilkołazka, and Andrzej Paszczyński, "Lignin Degradation: Microorganisms, Enzymes Involved, Genomes Analysis and Evolution," *FEMS Microbiology Reviews* 41 (2017): 941–62. doi: 10.1093/femsre/fux049.

15. Andrew C. Scott and Ian J. Glasspool, "The Diversification of Paleozoic Fire Systems and Fluctuations in Atmospheric Oxygen Concentration," *Proceedings of the National Academy of Sciences* 103 (2006): 10861–65. doi: 10.1073/pnas.0604090103.

16. Wenkun Qie, Thomas J. Algeo, Genming Luo, and Achim Herrmann, "Global Events of the Late Paleozoic (Early Devonian to Middle Permian): A Review,"

Palaeogeography, Palaeoclimatology, Palaeoecology 531 (2019): 109259. https://doi.org/10.1016/j.palaeo.2019.109259.

17. Friedlingstein et al., "Carbon Budget."

18. Johannes Lehmann and Markus Kleber, "The Contentious Nature of Soil Organic Matter," *Nature* 528 (2015): 60–68. doi: 10.1038/nature16069.

19. Zheng Shi, Steven D. Allison, Yujie He, Paul A. Levine, Alison M. Hoyt, Jeffrey Beem-Miller, Qing Zhu, et al., "The Age Distribution of Global Soil Carbon Inferred from Radiocarbon Measurements," *Nature Geoscience* 13 (2020): 555–59. doi: 10.1038/s41561-020-0596-z.

20. https://www.nobelprize.org/prizes/medicine/1952/waksman/biographical/, retrieved February 23, 2022.

21. Philippe C. Baveye and Michelle Wander, "The (Bio)Chemistry of Soil Humus and Humic Substances: Why Is the 'New View' Still Considered Novel after More Than 80 Years?" *Frontiers in Environmental Science* 7 (2019). doi: 10.3389/fenvs.2019.00027.

22. Selman A. Waksman, *Humus; Origin, Chemical Composition, and Importance in Nature* (Baltimore: The Williams & Wilkins Company, 1936).

23. Waksman, *Humus*, p. xi.

24. Baveye and Wander, "Soil Humus."

25. Laureano A. Gherardi and Osvaldo E. Sala, "Global Patterns and Climatic Controls of Belowground Net Carbon Fixation," *Proceedings of the National Academy of Sciences* 117 (2020): 20038–43. doi: 10.1073/pnas.2006715117.

26. Michael W. I. Schmidt, Margaret S. Torn, Samuel Abiven, Thorsten Dittmar, Georg Guggenberger, Ivan A. Janssens, Markus Kleber, et al., "Persistence of Soil Organic Matter as an Ecosystem Property," *Nature* 478 (2011): 49–56. doi: 10.1038/nature10386.

27. Anja Miltner, Petra Bombach, Burkhard Schmidt-Brücken, and Matthias Kästner, "SOM Genesis: Microbial Biomass as a Significant Source," *Biogeochemistry* 111 (2012): 41–55. doi: 10.1007/s10533-011-9658-z.

28. Baorong Wang, Shaoshan An, Chao Liang, Yang Liu, and Yakov Kuzyakov, "Microbial Necromass as the Source of Soil Organic Carbon in Global Ecosystems," *Soil Biology and Biochemistry* 162 (2021): 108422. https://doi.org/10.1016/j.soilbio.2021.108422.

29. Sokol et al., "Soil Microbiome."

30. Tessa Camenzind, Kyle Mason-Jones, India Mansour, Matthias C. Rillig, and Johannes Lehmann, "Formation of Necromass-Derived Soil Organic Carbon Determined by Microbial Death Pathways," *Nature Geoscience* 16 (2023): 115–22. doi: 10.1038/s41561-022-01100-3.

31. R. T. Watson, H. Rodhe, H. Oeschger, and U. Siegenthaler, "Greenhouse Gases and Aerosols," in *Climate Change: The IPCC Scientific Assessment*, ed. J. T. Houghton, G. J. Jenkins, and J. J. Ephraums (Cambridge: Cambridge University Press, 1990), pp. 1–40.

32. Anthony P. Walker, Martin G. De Kauwe, Ana Bastos, Soumaya Belmecheri, Katerina Georgiou, Ralph F. Keeling, Sean M. McMahon, et al., "Integrating the Evidence for a Terrestrial Carbon Sink Caused by Increasing Atmospheric CO_2," *New Phytologist* 229 (2021): 2413–45. https://doi.org/10.1111/nph.16866.

33. M. Fernández-Martínez, S. Vicca, I. A. Janssens, P. Ciais, M. Obersteiner, M. Bartrons, J. Sardans, et al., "Atmospheric Deposition, CO_2, and Change in the Land Carbon Sink," *Scientific Reports* 7 (2017): 9632. doi: 10.1038/s41598-017-08755-8.

34. J. E. Campbell, J. A. Berry, U. Seibt, S. J. Smith, S. A. Montzka, T. Launois, S. Belviso, et al., "Large Historical Growth in Global Terrestrial Gross Primary Production," *Nature* 544 (2017): 84–87. doi: 10.1038/nature22030.

35. Chi Chen, William J. Riley, I. Colin Prentice, and Trevor F. Keenan, "CO_2 Fertilization of Terrestrial Photosynthesis Inferred from Site to Global Scales," *Proceedings of the National Academy of Sciences* 119 (2022): e2115627119. doi:10.1073/pnas.2115627119.

36. Elizabeth A. Ainsworth and Stephen P. Long, "30 Years of Free-Air Carbon Dioxide Enrichment (FACE): What Have We Learned About Future Crop Productivity and Its Potential for Adaptation?" *Global Change Biology* 27 (2021): 27–49. https://doi.org/10.1111/gcb.15375.

37. César Terrer, Sara Vicca, Bruce A. Hungate, Richard P. Phillips, and I. Colin Prentice, "Mycorrhizal Association as a Primary Control of the CO_2 Fertilization Effect," *Science* 353 (2016): 72–74. doi:10.1126/science.aaf4610.

38. Mark A. Anthony, Thomas W. Crowther, Sietse van der Linde, Laura M. Suz, Martin I. Bidartondo, Filipa Cox, Marcus Schaub, et al., "Forest Tree Growth Is Linked to Mycorrhizal Fungal Composition and Function across Europe," *The ISME Journal* 16 (2022): 1327–36. doi: 10.1038/s41396-021-01159-7.

39. Ben Bond-Lamberty and Allison Thomson, "Temperature-Associated Increases in the Global Soil Respiration Record," *Nature* 464 (2010): 579–82. doi: 10.1038/nature08930.

40. Jiesi Lei, Xue Guo, Yufei Zeng, Jizhong Zhou, Qun Gao, and Yunfeng Yang, "Temporal Changes in Global Soil Respiration Since 1987," *Nature Communications* 12 (2021): 403. doi: 10.1038/s41467-020-20616-z.

41. Pablo García-Palacios, Thomas W. Crowther, Marina Dacal, Iain P. Hartley, Sabine Reinsch, Riikka Rinnan, Johannes Rousk, et al., "Evidence for Large Microbial-Mediated Losses of Soil Carbon under Anthropogenic Warming," *Nature Reviews Earth & Environment* 2 (2021): 507–17. doi: 10.1038/s43017-021-00178-4.

42. Xin Wang, Lingli Liu, Shilong Piao, Ivan A. Janssens, Jianwu Tang, Weixing Liu, Yonggang Chi, et al., "Soil Respiration under Climate Warming: Differential Response of Heterotrophic and Autotrophic Respiration," *Global Change Biology* 20 (2014): 3229–37. https://doi.org/10.1111/gcb.12620.

43. Ben Bond-Lamberty, Vanessa L. Bailey, Min Chen, Christopher M. Gough, and Rodrigo Vargas, "Globally Rising Soil Heterotrophic Respiration over Recent Decades," *Nature* 560 (2018): 80–83. doi: 10.1038/s41586-018-0358-x.

44. Dilip G. T. Naidu and Sumanta Bagchi, "Greening of the Earth Does Not Compensate for Rising Soil Heterotrophic Respiration under Climate Change," *Global Change Biology* 27 (2021): 2029–38. https://doi.org/10.1111/gcb.15531.

45. Lei Cheng, Naifang Zhang, Mengting Yuan, Jing Xiao, Yujia Qin, Ye Deng, Qichao Tu, et al., "Warming Enhances Old Organic Carbon Decomposition through Altering Functional Microbial Communities," *The ISME Journal* 11 (2017): 1825–35. doi: 10.1038/ismej.2017.48.

46. Sharon L. Smith, H. Brendan O'Neill, Ketil Isaksen, Jeannette Noetzli, and Vladimir E. Romanovsky, "The Changing Thermal State of Permafrost," *Nature Reviews Earth & Environment* 3 (2022): 10–23. doi: 10.1038/s43017-021-00240-1.

47. Hal Bernton, "As Alaska Permafrost Melts, Roads Sink, Bridges Tilt and Greenhouse Gases Escape," *Anchorage Daily News*, December 19, 2019, https://www.adn.com/ala ska-news/weather/2019/12/17/as-alaska-permafrost-melts-roads-sink-bridges-tilt-and-greenhouse-gases-escape/, accessed March 3, 2022.

48. Jan Hjort, Olli Karjalainen, Juha Aalto, Sebastian Westermann, Vladimir E. Romanovsky, Frederick E. Nelson, Bernd Etzelmüller, et al., "Degrading Permafrost Puts Arctic Infrastructure at Risk by Mid-Century," *Nature Communications* 9 (2018): 5147. doi: 10.1038/s41467-018-07557-4.

49. Jan Hjort, Dmitry Streletskiy, Guy Doré, Qingbai Wu, Kevin Bjella, and Miska Luoto, "Impacts of Permafrost Degradation on Infrastructure," *Nature Reviews Earth & Environment* 3 (2022): 24–38. doi: 10.1038/s43017-021-00247-8.

50. Zhi-Ping Zhong, Funing Tian, Simon Roux, M. Consuelo Gazitúa, Natalie E. Solonenko, Yueh-Fen Li, Mary E. Davis, et al., "Glacier Ice Archives Nearly 15,000-Year-Old Microbes and Phages," *Microbiome* 9 (2021): 160. doi: 10.1186/s40168-021-01106-w.

51. Kevin Schaefer, Hugues Lantuit, Vladimir E. Romanovsky, Edward A. G. Schuur, and Ronald Witt, "The Impact of the Permafrost Carbon Feedback on Global Climate," *Environmental Research Letters* 9 (2014): 085003. doi: 10.1088/1748-9326/9/8/085003.

52. Edward A. G. Schuur,, Rosvel Bracho, Gerardo Celis, E. Fay Belshe, Chris Ebert, Justin Ledman, Marguerite Mauritz, et al., "Tundra Underlain by Thawing Permafrost Persistently Emits Carbon to the Atmosphere over 15 Years of Measurements," *Journal of Geophysical Research: Biogeosciences* 126 (2021): e2020JG006044. https://doi.org/10.1029/2020JG006044.

53. Merritt R. Turetsky, Brian Benscoter, Susan Page, Guillermo Rein, Guido R. van der Werf, and Adam Watts, "Global Vulnerability of Peatlands to Fire and Carbon Loss," *Nature Geoscience* 8 (2015): 11–14. doi: 10.1038/ngeo2325.

54. Jonathan E. Nichols and Dorothy M. Peteet, "Rapid Expansion of Northern Peatlands and Doubled Estimate of Carbon Storage," *Nature Geoscience* 12 (2019): 917–21. doi: 10.1038/s41561-019-0454-z.

55. Turetsky et al., "Global Vulnerability."

Chapter 4

1. Robert K. Musil, "There Must Be More to Love Than Death: A Conversation with Kurt Vonnegut," *The Nation* 231 (1980): 128–32.

2. Alexander Rosu-Finsen, Michael B. Davies, Alfred Amon, Han Wu, Andrea Sella, Angelos Michaelides, and Christoph G. Salzmann, "Medium-Density Amorphous Ice," *Science* 379 (2023): 474–78. doi:10.1126/science.abq2105.

3. Dean Roemmich, John Church, John Gilson, Didier Monselesan, Philip Sutton, and Susan Wijffels, "Unabated Planetary Warming and Its Ocean Structure Since 2006," *Nature Climate Change* 5 (2015): 240–45. doi: 10.1038/nclimate2513.

4. IPCC, "Summary for Policymakers," in *Climate Change 2021: The Physical Science Basis. Contribution of Working Group I to the Sixth Assessment Report of the Intergovernmental Panel on Climate Change*, ed. V. Masson-Delmotte, P. Zhai, A. Pirani, S. L. Connors, C. Péan, S. Berger, N. Caud, Y. Chen, L. Goldfarb, M. I. Gomis, M. Huang, K. Leitzell, E. Lonnoy, J. B. R. Matthews, T. K. Maycock, T. Waterfield, O. Yelekçi, R. Yu, and B. Zhou (Cambridge: Cambridge University Press, 2021), pp. 3–32.

5. T. Wernberg, D. A. Smale, T. L. Frölicher, and A. J. P. Smith, "Sciencebrief Review: Climate Change Increases Marine Heatwaves Harming Marine Ecosystems," ed. P. Liss, C. Le Quéré, and P. Forster, *Critical Issues in Climate Change Science* (2021), https://doi.org/10.5281/zenodo.5596820.

6. A. Capotondi, M. Newman, T. Xu, and E. Di Lorenzo, "An Optimal Precursor of Northeast Pacific Marine Heatwaves and Central Pacific El Niño Events," *Geophysical Research Letters* 49 (2022): e2021GL097350. https://doi.org/10.1029/2021GL097350.

7. Wallace S. Broecker, *The Great Ocean Conveyor: Discovering the Trigger for Abrupt Climate Change* (Princeton, NJ: Princeton University Press, 2010).

8. Jorge Louis Sarmiento and Nicolas Gruber, *Ocean Biogeochemical Dynamics* (Princeton, NJ: Princeton University Press, 2006).

9. Jamie D. Wilson, Oliver Andrews, Anna Katavouta, Francisco de Melo Viríssimo, Ros M. Death, Markus Adloff, Chelsey A. Baker, et al., "The Biological Carbon Pump in CMIP6 Models: 21st Century Trends and Uncertainties," *Proceedings of the National Academy of Sciences* 119 (2022): e2204369119. doi:10.1073/pnas.2204369119.

10. Michael Nowicki, Tim DeVries, and David A. Siegel, "Quantifying the Carbon Export and Sequestration Pathways of the Ocean's Biological Carbon Pump," *Global Biogeochemical Cycles* 36 (2022): e2021GB007083. https://doi.org/10.1029/2021GB007083.

11. Margaret Deacon, *Scientists and the Sea, 1650–1900: A Study of Marine Science*, 2nd ed. (Aldershot, UK: Ashgate, 1997).

12. Alexander Agassiz, *A Contribution to American Thalassography: Three Cruises of the United States Coast and Geodetic Survey Steamer "Blake," in the Gulf of Mexico, in the Caribbean Sea, and Along the Atlantic Coast of the United States, from 1877 to 1880*, Bulletin of the Museum of Comparative Zoology, vol. 14–15 (Boston: Houghton, Mifflin and Co., 1888).

13. The publications mentioning "rain of detritus" all say Agassiz came up with the phrase, and if they cite anything, it's Agassiz (1888), but I did not find the phrase in that publication.

14. Linda J. Lear, *Rachel Carson: Witness for Nature* (Boston: Mariner Books, 2009).

15. Lear, *Rachel Carson*, p. 168.

16. Rachel Carson, *The Sea around Us*, rev ed. (New York: Oxford University Press, 1961), p. 71

17. Mary Silver, "Marine Snow: A Brief Historical Sketch," *Limnology and Oceanography Bulletin* 24 (2015): 5–10. https://doi.org/10.1002/lob.10005.

18. Tokimi Tsujita, "Studies on Naturally Occurring Suspended Organic Matter in the Waters Adjacent to Japan (III): On a Process of Organization of Planktogenic Organic Matter as Examined by the Electron Microscope," *Journal of the Oceanographic Society of Japan* 11 (1955): 199–203. doi: 10.5928/kaiyou1942.11.199.

19. Noboru Suzuki and Kenji Kato, "Studies on Suspended Materials Marine Snow in the Sea. Part I. Sources of Marine Snow," *Bulletin of the Faculty of Fisheries Hokkaido University* 4 (1953): 132–37. http://hdl.handle.net/2115/22805.

20. Nowicki et al., "Quantifying Carbon Export."

21. Dennis A. Hansell, "Recalcitrant Dissolved Organic Carbon Fractions," *Annual Review of Marine Science* 5 (2013): 421–45. doi: 10.1146/annurev-marine-120710-100757.

22. Mary Ann Moran, Frank X. Ferrer-González, He Fu, Brent Nowinski, Malin Olofsson, McKenzie A. Powers, Jeremy E. Schreier, et al., "The Ocean's Labile DOC Supply Chain," *Limnology and Oceanography* 67 (2022): 1007–21. https://doi.org/10.1002/lno.12053.

23. Peter J. le B. Williams and Hugh W. Ducklow, "The Microbial Loop Concept: A History, 1930–1974," *Journal of Marine Research* 77 (2019): 23–81.

24. Moran et al., "DOC Supply Chain."

25. Steven W. Wilhelm and Curtis A. Suttle, "Viruses and Nutrient Cycles in the Sea: Viruses Play Critical Roles in the Structure and Function of Aquatic Food Webs," *BioScience* 49 (1999): 781–88. doi: 10.2307/1313569.

26. Email to the author, August 18, 2022.

27. Hansell, "Recalcitrant DOC."

28. Peter M. Williams and Ellen R. M. Druffel, "Radiocarbon in Dissolved Organic Matter in the Central North Pacific Ocean," *Nature* 330 (1987): 246–48. doi: 10.1038/330246a0.

29. Christopher L. Follett, Daniel J. Repeta, Daniel H. Rothman, Li Xu, and Chiara Santinelli, "Hidden Cycle of Dissolved Organic Carbon in the Deep Ocean," *Proceedings of the National Academy of Sciences* 111 (2014): 16706–11. doi:10.1073/pnas.1407445111.

30. Hansell, "Recalcitrant DOC."

31. Nowicki et al., "Quantifying Carbon Export."

32. Nianzhi Jiao, Gerhard J. Herndl, Dennis A. Hansell, Ronald Benner, Gerhard Kattner, Steven W. Wilhelm, David L. Kirchman, et al., "Microbial Production of Recalcitrant Dissolved Organic Matter: Long-Term Carbon Storage in the Global Ocean," *Nature Reviews Microbiology* 8 (2010): 593–99. doi: 10.1038/nrmicro2386.

33. S. K. Gulev, P. W. Thorne, J. Ahn, F. J. Dentener, C. M. Domingues, S. Gerland, D. Gong, et al., "Changing State of the Climate System," in *Climate Change 2021: The Physical Science Basis. Contribution of Working Group I to the Sixth Assessment Report of the Intergovernmental Panel on Climate Change*, ed. V. Masson-Delmotte, P. Zhai, A. Pirani, S. L. Connors, C. Péan, S. Berger, N. Caud, Y. Chen, L. Goldfarb, M. I. Gomis, M. Huang, K. Leitzell, E. Lonnoy, J. B. R. Matthews, T. K. Maycock, T. Waterfield, O. Yelekçi, R. Yu, and B. Zhou (Cambridge: Cambridge University Press, 2021), pp. 287–422.

34. Hansell, "Recalcitrant DOC."

35. Selman A. Waksman, "On the Distribution of Organic Matter in the Sea Bottom and the Chemical Nature and Origin of Marine Humus," *Soil Science* 36 (1933): 125–47.

36. Ronald Benner and Rainer M. W. Amon, "The Size-Reactivity Continuum of Major Bioelements in the Ocean," *Annual Review of Marine Science* 7 (2015): 185–205. doi: 10.1146/annurev-marine-010213-135126.

37. N. Hertkorn, R. Benner, M. Frommberger, P. Schmitt-Kopplin, M. Witt, K. Kaiser, A. Kettrup, et al., "Characterization of a Major Refractory Component of Marine Dissolved Organic Matter," *Geochimica et Cosmochimica Acta* 70 (2006): 2990–3010. doi: 10.1016/j.gca.2006.03.021.

38. D. Scott Snyder, Bianca Brahamsha, Parastoo Azadi, and Brian Palenik, "Structure of Compositionally Simple Lipopolysaccharide from Marine *Synechococcus*," *Journal of Bacteriology* 191 (2009): 5499–509. doi:10.1128/JB.00121-09.

39. Hiroshi Ogawa, Yukio Amagai, Isao Koike, Karl Kaiser, and Ronald Benner, "Production of Refractory Dissolved Organic Matter by Bacteria," *Science* 292 (2001): 917–20. doi:10.1126/science.1057627.

40. The "300 times more resilient" estimate is the ratio of the lifetime of the labile compounds (a day or two) to the minimum residence time of the bacteria-produced recalcitrant DOC (1.5 years).

41. Ronald Benner and Gerhard J. Herndl, "Bacterially Derived Dissolved Organic Matter in the Microbial Carbon Pump," in *Microbial Carbon Pump in the Ocean*, ed. Nianzhi Jiao, Farooq Azam, and Sean Sanders (Washington, DC: Science/AAAS, 2011), pp. 46–48. doi: 10.1126/science.opms.sb0001

42. Thorsten Dittmar, Sinikka T. Lennartz, Hagen Buck-Wiese, Dennis A. Hansell, Chiara Santinelli, Chiara Vanni, Bernd Blasius, et al., "Enigmatic Persistence of Dissolved Organic Matter in the Ocean," *Nature Reviews Earth & Environment* 2 (2021): 570–83. doi: 10.1038/s43017-021-00183-7.

43. P. Friedlingstein, M. O'Sullivan, M. W. Jones, R. M. Andrew, L. Gregor, J. Hauck, C. Le Quéré, et al. "Global Carbon Budget 2022." *Earth System Science Data* 14 (2022): 4811–900. doi: 10.5194/essd-14-4811-2022.

44. David A. Hutchins and Feixue Fu, "Microorganisms and Ocean Global Change," *Nature Microbiology* 2 (2017): 17058. doi: 10.1038/nmicrobiol.2017.58.

45. Daniel G. Boyce, Marlon R. Lewis, and Boris Worm, "Global Phytoplankton Decline over the Past Century," *Nature* 466 (2010): 591–96. doi: 10.1038/nature09268.

46. David A. Siegel and Bryan A. Franz, "Century of Phytoplankton Change," *Nature* 466 (2010): 569–71. doi: 10.1038/466569a.

47. Sergio R. Signorini, Bryan A. Franz, and Charles R. McClain, "Chlorophyll Variability in the Oligotrophic Gyres: Mechanisms, Seasonality and Trends," *Frontiers in Marine Science* 2 (2015). doi: 10.3389/fmars.2015.00001.

48. Andy Ridgwell and Sandra Arndt, "Why Dissolved Organics Matter: DOC in Ancient Oceans and Past Climate Change," in *Biogeochemistry of Marine Dissolved Organic Matter*, 2nd ed., ed. Dennis A. Hansell and Craig A. Carlson (Boston: Academic Press, 2015), pp. 1–20.

49. Urs Siegenthaler, Thomas F. Stocker, Eric Monnin, Dieter Lüthi, Jakob Schwander, Bernhard Stauffer, Dominique Raynaud, et al., "Stable Carbon Cycle—Climate Relationship during the Late Pleistocene," *Science* 310 (2005): 1313–17. doi:10.1126/science.1120130.

50. Ji-Woong Yang, Margaux Brandon, Amaëlle Landais, Stéphanie Duchamp-Alphonse, Thomas Blunier, Frédéric Prié, and Thomas Extier, "Global Biosphere Primary Productivity Changes during the Past Eight Glacial Cycles," *Science* 375 (2022): 1145–51. doi: 10.1126/science.abj8826.

51. Daniel M. Sigman, François Fripiat, Anja S. Studer, Preston C. Kemeny, Alfredo Martínez-García, Mathis P. Hain, Xuyuan Ai, et al., "The Southern Ocean during the Ice Ages: A Review of the Antarctic Surface Isolation Hypothesis, with Comparison to the North Pacific," *Quaternary Science Reviews* 254 (2021): 106732. https://doi.org/10.1016/j.quascirev.2020.106732.

52. John H. Martin, "Glacial-Interglacial CO_2 Change: The Iron Hypothesis," *Paleoceanography* 5 (1990): 1–13. https://doi.org/10.1029/PA005i001p00001.

53. Daniel M. Sigman and Edward A. Boyle, "Glacial/Interglacial Variations in Atmospheric Carbon Dioxide," *Nature* 407 (2000): 859–69. doi: 10.1038/35038000.

54. Natalie Mahowald, Karen Kohfeld, Margaret Hansson, Yves Balkanski, Sandy P. Harrison, I. Colin Prentice, Michael Schulz, et al., "Dust Sources and Deposition during the Last Glacial Maximum and Current Climate: A Comparison of Model Results with Paleodata from Ice Cores and Marine Sediments," *Journal of Geophysical Research: Atmospheres* 104 (1999): 15895–916. https://doi.org/10.1029/1999JD900084.

55. Wilson et al., "Biological Carbon Pump."

56. Andrew J. Irwin and Matthew J. Oliver, "Are Ocean Deserts Getting Larger?" *Geophysical Research Letters* 36 (2009). https://doi.org/10.1029/2009GL039883.

57. Hutchins and Fu, "Microorganisms."

58. Sunke Schmidtko, Lothar Stramma, and Martin Visbeck, "Decline in Global Oceanic Oxygen Content during the Past Five Decades," *Nature* 542 (2017): 335–39. doi: 10.1038/nature21399.

59. Takamitsu Ito, Shoshiro Minobe, Matthew C. Long, and Curtis Deutsch, "Upper Ocean O_2 Trends: 1958–2015," *Geophysical Research Letters* 44 (2017): 4214–23. 10.1002/2017GL073613.

60. Schmidtko et al., "Global Oceanic Oxygen."

61. Keith J. Laidler, "The Development of the Arrhenius Equation," *Journal of Chemical Education* 61 (1984): 494–98. doi: 10.1021/ed061p494.

62. Eric A. Davidson and Ivan A. Janssens, "Temperature Sensitivity of Soil Carbon Decomposition and Feedbacks to Climate Change," *Nature* 440 (2006): 165–73. doi: 10.1038/nature04514.

63. Pablo García-Palacios, Thomas W. Crowther, Marina Dacal, Iain P. Hartley, Sabine Reinsch, Riikka Rinnan, Johannes Rousk, et al., "Evidence for Large Microbial-Mediated Losses of Soil Carbon under Anthropogenic Warming," *Nature Reviews Earth & Environment* 2 (2021): 507–17. doi: 10.1038/s43017-021-00178-4.

64. Peter G. Brewer, "The Molecular Basis for Understanding the Impacts of Ocean Warming," *Reviews of Geophysics* 57 (2019): 1112–23. https://doi.org/10.1029/2018RG000620.

65. Christian Lønborg, Xosé A. Álvarez–Salgado, Robert T. Letscher, and Dennis A. Hansell, "Large Stimulation of Recalcitrant Dissolved Organic Carbon Degradation

by Increasing Ocean Temperatures," *Frontiers in Marine Science* 4 (2018). doi: 10.3389/fmars.2017.00436.

66. Joachim Segschneider and Jørgen Bendtsen, "Temperature-Dependent Remineralization in a Warming Ocean Increases Surface pCO$_2$ through Changes in Marine Ecosystem Composition," *Global Biogeochemical Cycles* 27 (2013): 1214–25. https://doi.org/10.1002/2013GB004684.

67. Wilson et al., "Biological Carbon Pump."

68. P. Forster, T. Storelvmo, K. Armour, W. Collins, J.-L. Dufresne, D. Frame, D. J. Lunt, et al., "The Earth's Energy Budget, Climate Feedbacks, and Climate Sensitivity," in *Climate Change 2021: The Physical Science Basis. Contribution of Working Group I to the Sixth Assessment Report of the Intergovernmental Panel on Climate Change*, ed. V. Masson-Delmotte, P. Zhai, A. Pirani, S. L. Connors, C. Péan , S. Berger, N. Caud, Y. Chen, L. Goldfarb, M. I. Gomis, M. Huang, K. Leitzell, E. Lonnoy, J. B. R. Matthews, T. K. Maycock, T. Waterfield, O. Yelekçi, R. Yu, and B. Zhou (Cambridge: Cambridge University Press, 2021), pp. 923–1054.

Chapter 5

1. Elisabeth Crawford, *Arrhenius: From Ionic Theory to Greenhouse Effect* (Canton, MA: Science History Publications, 1996).

2. Andrei G. Lapenis, "Arrhenius and the Intergovernmental Panel on Climate Change," *Eos* 79 (1998): 271. https://doi.org/10.1029/98EO00206.

3. D. Chen, M. Rojas, B. H. Samset, K. Cobb, A. Diongue Niang, P. Edwards, S. Emori, et al., "Framing, Context, and Methods," in *Climate Change 2021: The Physical Science Basis. Contribution of Working Group I to the Sixth Assessment Report of the Intergovernmental Panel on Climate Change*, ed. V. Masson-Delmotte, P. Zhai, A. Pirani, S. L. Connors, C. Péan , S. Berger, N. Caud, Y. Chen, L. Goldfarb, M. I. Gomis, M. Huang, K. Leitzell, E. Lonnoy, J. B. R. Matthews, T .K. Maycock, T. Waterfield, O. Yelekçi, R. Yu, and B. Zhou (Cambridge: Cambridge University Press, 2021), pp. 147–286; Chen et al., "Framing, Context, and Methods."

4. Svante Arrhenius, *Worlds in the Making: The Evolution of the Universe*, trans. H. Borns (London: Harper & Brothers, 1908), p. 61.

5. Arrhenius, *Worlds*, p. 63.

6. P. Forster, T. Storelvmo, K. Armour, W. Collins, J.-L. Dufresne, D. Frame, D. J. Lunt, et al., "The Earth's Energy Budget, Climate Feedbacks, and Climate Sensitivity," in *Climate Change 2021: The Physical Science Basis. Contribution of Working Group I to the Sixth Assessment Report of the Intergovernmental Panel on Climate Change*, ed. V. Masson-Delmotte, P. Zhai, A. Pirani, S. L. Connors, C. Péan , S. Berger, N. Caud, Y. Chen, L. Goldfarb, M. I. Gomis, M. Huang, K. Leitzell, E. Lonnoy, J. B. R. Matthews, T. K. Maycock, T. Waterfield, O. Yelekçi, R. Yu, and B. Zhou (Cambridge: Cambridge University Press, 2021), pp. 923–1054.

7. Mark D. Zelinka, Timothy A. Myers, Daniel T. McCoy, Stephen Po-Chedley, Peter M. Caldwell, Paulo Ceppi, Stephen A. Klein, et al., "Causes of Higher Climate Sensitivity

in CMIP6 Models," *Geophysical Research Letters* 47 (2020): e2019GL085782. https://doi.org/10.1029/2019GL085782.

8. R. R. Rogers and Man Kong Yau, *A Short Course in Cloud Physics*, 3rd ed., vol. 113, International Series in Natural Philosophy (Oxford: Butterworth-Heinemann, 1989).

9. James Lovelock, R. A. Cox, P. S. Liss, and P. Borrell, "A Geophysiologist's Thoughts on the Natural Sulphur Cycle," *Philosophical Transactions: Biological Sciences* 352 (1997): 143–47. https://www.jstor.org/stable/56556.

10. John Gribbin and Mary Gribbin, *He Knew He Was Right: The Irrepressible Life of James Lovelock and Gaia* (London: Allen Lane, 2009).

11. Matthew S. Savoca, "Chemoattraction to Dimethyl Sulfide Links the Sulfur, Iron, and Carbon Cycles in High-Latitude Oceans," *Biogeochemistry* 138 (2018): 1–21. doi: 10.1007/s10533-018-0433-2.

12. J. E. Lovelock, R. J. Maggs, and R. A. Rasmussen, "Atmospheric Dimethyl Sulphide and the Natural Sulphur Cycle," *Nature* 237 (1972): 452–53. doi: 10.1038/237452a0.

13. R. J. Charlson, J. E. Lovelock, M. O. Andreae, and S. G. Warren, "Oceanic Phytoplankton, Atmospheric Sulfur, Cloud Albedo and Climate," *Nature* 326 (1987): 655–61. https://doi.org/10.1038/326655a0.

14. Kirsten N. Fossum, Jurgita Ovadnevaite, Darius Ceburnis, Jana Preißler, Jefferson R. Snider, Ru-Jin Huang, Andreas Zuend, et al., "Sea-Spray Regulates Sulfate Cloud Droplet Activation over Oceans," *npj Climate and Atmospheric Science* 3 (2020). doi: 10.1038/s41612-020-0116-2.

15. Mary Ann Moran, Chris R. Reisch, Ronald P. Kiene, and William B. Whitman, "Genomic Insights into Bacterial DMSP Transformations," *Annual Review of Marine Science* 4 (2012): 523–42. doi: 10.1146/annurev-marine-120710-100827.

16. P. Fraser, "A Brief History of the Cape Grim Baseline Air Pollution Station," in *Baseline Atmospheric Program Australia 2005–2006*, ed. J. Cainey, N. Derek, and P. Krummel (Melbourne: Australian Bureau of Meteorology and CSIRO Marine and Atmospheric Research, 2007), pp. 1–6.

17. P. K. Quinn and T. S. Bates, "The Case against Climate Regulation Via Oceanic Phytoplankton Sulphur Emissions," *Nature* 480 (2011): 51–56. doi: 10.1038/nature10580.

18. Charlson et al., "Oceanic Phytoplankton," p. 655.

19. J. E. Lovelock, "Gaia as Seen through the Atmosphere," *Atmospheric Environment* 6 (1972): 579–80. https://doi.org/10.1016/0004-6981(72)90076-5.

20. James Lovelock, *Gaia: A New Look at Life on Earth* (Oxford: Oxford University Press, 1987), p. 10.

21. Gribbin and Gribbin, *He Knew*.

22. Gribbin and Gribbin, *He Knew*, p. 94.

23. Timothy M. Lenton, "Gaia and Natural Selection," *Nature* 394 (1998): 439–47. doi: 10.1038/28792.

24. https://www.c-span.org/video/?125856-1/greenhouse-effect, accessed May 3, 2022.

25. Andrew H. Knoll, "Lynn Margulis, 1938–2011," *Proceedings of the National Academy of Sciences* 109 (2012): 1022. doi: 10.1073/pnas.1120472109.

26. Tom A. Williams, Cymon J. Cox, Peter G. Foster, Gergely J. Szöllősi, and T. Martin Embley, "Phylogenomics Provides Robust Support for a Two-Domains Tree of Life," *Nature Ecology & Evolution* 4 (2020): 138–47. doi: 10.1038/s41559-019-1040-x.

27. Gribbin and Gribbin, *He Knew.*

28. James E. Lovelock and Lynn Margulis, "Atmospheric Homeostasis by and for the Biosphere: The Gaia Hypothesis," *Tellus* 26 (1974): 2–10. doi: 10.3402/tellusa. v26i1-2.9731.

29. Lynn Margulis and J. E. Lovelock, "Biological Modulation of the Earth's Atmosphere," *Icarus* 21 (1974): 471–89. https://doi.org/10.1016/0019-1035(74)90150-X.

30. Lovelock and Margulis, "Atmospheric Homeostasis," p. 8.

31. Toby Tyrrell, *On Gaia: A Critical Investigation of the Relationship between Life and Earth* (Princeton, NJ: Princeton University Press, 2013).

32. Lovelock, *Gaia*, p. 11.

33. The two time periods examined were 1974 through 1984 and from 2011 through 2021. The data are from a Web of Science analysis, on May 9, 2022.

34. Sergio M. Vallina and Rafel Simó, "Strong Relationship between DMS and the Solar Radiation Dose over the Global Surface Ocean," *Science* 315 (2007): 506–08. doi: 10.1126/science.1133680.

35. Tamara K. Green, T. K. Green, and A. D. Hatton, "The CLAW Hypothesis: A New Perspective on the Role of Biogenic Sulphur in the Regulation of Global Climate," *Oceanography and Marine Biology: An Annual Review* 52 (2014): 315–35.

36. Shanlin Wang, Mathew Maltrud, Scott Elliott, Philip Cameron-Smith, and Alexandra Jonko, "Influence of Dimethyl Sulfide on the Carbon Cycle and Biological Production," *Biogeochemistry* 138 (2018): 49–68. doi: 10.1007/s10533-018-0430-5.

37. P. K. Quinn, D. J. Coffman, J. E. Johnson, L. M. Upchurch, and T. S. Bates, "Small Fraction of Marine Cloud Condensation Nuclei Made up of Sea Spray Aerosol," *Nature Geoscience* 10 (2017): 674–79. doi: 10.1038/ngeo3003.

38. Sarah D. Brooks and Daniel C.O. Thornton, "Marine Aerosols and Clouds," *Annual Review of Marine Science* 10 (2018): 289–313. doi: 10.1146/annurev-marine-121916-063148.

39. Zhujun Yu and Ying Li, "Marine Volatile Organic Compounds and Their Impacts on Marine Aerosol—A Review," *Science of The Total Environment* 768 (2021): 145054. https://doi.org/10.1016/j.scitotenv.2021.145054.

40. Quinn and Bates, "The Case Against."

41. Charlson et al., "Oceanic Phytoplankton," p. 660.

42. A. Lana, T. G. Bell, R. Simó, S. M. Vallina, J. Ballabrera-Poy, A. J. Kettle, J. Dachs, et al., "An Updated Climatology of Surface Dimethlysulfide Concentrations and Emission Fluxes in the Global Ocean," *Global Biogeochemical Cycles* 25 (2011): https://doi.org/10.1029/2010GB003850.

43. Duane C. Yoch, "Dimethylsulfoniopropionate: Its Sources, Role in the Marine Food Web, and Biological Degradation to Dimethylsulfide," *Applied and Environmental Microbiology* 68 (2002): 5804–15. doi:10.1128/AEM.68.12.5804-5815.2002.

44. Ki-Tae Park, Kitack Lee, Tae-Wook Kim, Young Jun Yoon, Eun-Ho Jang, Sehyun Jang, Bang-Yong Lee, et al., "Atmospheric DMS in the Arctic Ocean and Its Relation to

Phytoplankton Biomass," *Global Biogeochemical Cycles* 32 (2018): 351–59. https://doi.org/10.1002/2017GB005805.

45. C. Nissen and M. Vogt, "Factors Controlling the Competition between *Phaeocystis* and Diatoms in the Southern Ocean and Implications for Carbon Export Fluxes," *Biogeosciences* 18 (2021): 251–83. doi: 10.5194/bg-18-251-2021.

46. Quinn and Bates, "The Case Against," p. 55.

47. J. Bock, M. Michou, P. Nabat, M. Abe, J. P. Mulcahy, D. J. L. Olivié, J. Schwinger, et al., "Evaluation of Ocean Dimethylsulfide Concentration and Emission in CMIP6 Models," *Biogeosciences* 18 (2021): 3823–60. doi: 10.5194/bg-18-3823-2021.

48. Paul G. Falkowski, Yongseung Kim, Zbigniew Kolber, Cara Wilson, Creighton Wirick, and Robert Cess, "Natural Versus Anthropogenic Factors Affecting Low-Level Cloud Albedo over the North Atlantic," *Science* 256 (1992): 1311–13. doi: 10.1126/science.256.5061.1311.

49. Olaf Krüger and Hartmut Graßl, "Southern Ocean Phytoplankton Increases Cloud Albedo and Reduces Precipitation," *Geophysical Research Letters* 38 (2011): L08809. https://doi.org/10.1029/2011GL047116.

50. Karine Sellegri, Alessia Nicosia, Evelyn Freney, Julia Uitz, Melilotus Thyssen, Gérald Grégori, Anja Engel, et al., "Surface Ocean Microbiota Determine Cloud Precursors," *Scientific Reports* 11 (2021): 281. doi: 10.1038/s41598-020-78097-5.

51. Tyrrell, *On Gaia*, p. 208.

Chapter 6

1. https://en.wikipedia.org/wiki/White_Cliffs_of_Dover, accessed February 28, 2023.

2. Thomas Henry Huxley, *On a Piece of Chalk* (New York: Scribner, 1967).

3. Huxley, *Chalk*, p. 23.

4. Huxley, *Chalk*, p. 27.

5. Fanny M. Monteiro, Lennart T. Bach, Colin Brownlee, Paul Bown, Rosalind E. M. Rickaby, Alex J. Poulton, Toby Tyrrell, et al., "Why Marine Phytoplankton Calcify," *Science Advances* 2 (2016): e1501822. doi: 10.1126/sciadv.1501822.

6. Jorge Louis Sarmiento and Nicolas Gruber, *Ocean Biogeochemical Dynamics* (Princeton, NJ: Princeton University Press, 2006).

7. William W. Hay and Eloise Zakevich, "Cesare Emiliani (1922–1995): The Founder of Paleoceanography," *International Microbiology* 2 (1999): 52–54.

8. J. A. Raven and K. Crawfurd, "Environmental Controls on Coccolithophore Calcification," *Marine Ecology Progress Series* 470 (2012): 137–66. doi: 10.3354/meps09993.

9. Jason Hopkins, Stephanie A. Henson, Stuart C. Painter, Toby Tyrrell, and Alex J. Poulton, "Phenological Characteristics of Global Coccolithophore Blooms," *Global Biogeochemical Cycles* 29 (2015): 239–53. https://doi.org/10.1002/2014GB004919.

10. Sam Boggs, *Principles of Sedimentology and Stratigraphy*, 5th ed. (Upper Saddle River, NJ: Pearson Prentice Hall, 2012).

11. Sarmiento and Gruber, *Biogeochemical Dynamics.*

12. John McPhee, *Annals of the Former World*, 1st ed. (New York: Farrar, Straus and Giroux, 1998), p. 29.

13. McPhee, *Annals*, p. 29.

14. M. J. S. Rudwick, *Earth's Deep History: How It Was Discovered and Why It Matters* (Chicago: The University of Chicago Press, 2014).

15. Jan Pawlowski, Maria Holzmann, Cédric Berney, José Fahrni, Andrew J. Gooday, Tomas Cedhagen, Andrea Habura, et al., "The Evolution of Early Foraminifera," *Proceedings of the National Academy of Sciences* 100 (2003): 11494–98. doi: 10.1073/pnas.2035132100.

16. M. Débora Iglesias-Rodríguez, Christopher W. Brown, Scott C. Doney, Joan Kleypas, Dorota Kolber, Zbigniew Kolber, Paul K. Hayes, et al., "Representing Key Phytoplankton Functional Groups in Ocean Carbon Cycle Models: Coccolithophorids," *Global Biogeochemical Cycles* 16 (2002): 1100. https://doi.org/10.1029/2001GB001454.

17. Boggs, *Sedimentology.*

18. Judith A. McKenzie and Crisogono Vasconcelos, "Dolomite Mountains and the Origin of the Dolomite Rock of Which They Mainly Consist: Historical Developments and New Perspectives," *Sedimentology* 56 (2009): 205–19. https://doi.org/10.1111/j.1365-3091.2008.01027.x.

19. https://www.german-way.com/notable-people/featured-bios/german-geographer-ferdinand-von-richthofen-and-the-silk-road/, accessed July 1, 2022.

20. Ute Wardenga, "Ferdinand Von Richthofen and the Development of German Geography," *Die Erde* 138 (2007): 313–32.

21. McKenzie and Vasconcelos, "Dolomite Mountains."

22. Sara B. Pruss, Martha L. Slaymaker, Emily F. Smith, Andrey Yu Zhuravlev, and David A. Fike, "Cambrian Reefs in the Lower Poleta Formation: A New Occurrence of a Thick Archaeocyathan Reef near Gold Point, Nevada, USA," *Facies* 67 (2021): 14. doi: 10.1007/s10347-021-00623-2.

23. Jeana L. Drake, Tali Mass, Jarosław Stolarski, Stanislas Von Euw, Bas van de Schootbrugge, and Paul G. Falkowski, "How Corals Made Rocks through the Ages," *Global Change Biology* 26 (2020): 31–53. https://doi.org/10.1111/gcb.14912.

24. Andrew C. Baker, Peter W. Glynn, and Bernhard Riegl, "Climate Change and Coral Reef Bleaching: An Ecological Assessment of Long-Term Impacts, Recovery Trends and Future Outlook," *Estuarine, Coastal and Shelf Science* 80 (2008): 435–71. https://doi.org/10.1016/j.ecss.2008.09.003.

25. Terry P. Hughes, Kristen D. Anderson, Sean R. Connolly, Scott F. Heron, James T. Kerry, Janice M. Lough, Andrew H. Baird, et al., "Spatial and Temporal Patterns of Mass Bleaching of Corals in the Anthropocene," *Science* 359 (2018): 80–83. doi: 10.1126/science.aan8048.

26. Todd C. LaJeunesse, John Everett Parkinson, Paul W. Gabrielson, Hae Jin Jeong, James Davis Reimer, Christian R. Voolstra, and Scott R. Santos, "Systematic Revision of Symbiodiniaceae Highlights the Antiquity and Diversity of Coral Endosymbionts," *Current Biology* 28 (2018): 2570–80. https://doi.org/10.1016/j.cub.2018.07.008.

27. Ray Berkelmans and Madeleine J. H. van Oppen, "The Role of Zooxanthellae in the Thermal Tolerance of Corals: A 'Nugget of Hope' for Coral Reefs in an Era of Climate Change," *Proceedings of the Royal Society B: Biological Sciences* 273 (2006): 2305–12. doi: 10.1098/rspb.2006.3567.

28. LaJeunesse et al., "Systematic Revision."

29. Scott C. Doney, D. Shallin Busch, Sarah R. Cooley, and Kristy J. Kroeker, "The Impacts of Ocean Acidification on Marine Ecosystems and Reliant Human Communities," *Annual Review of Environment and Resources* 45 (2020): 83–112. doi: 10.1146/annurev-environ-012320-083019.

30. Bärbel Hönisch, Andy Ridgwell, Daniela N. Schmidt, Ellen Thomas, Samantha J. Gibbs, Appy Sluijs, Richard Zeebe, et al., "The Geological Record of Ocean Acidification," *Science* 335 (2012): 1058–63. doi: 10.1126/science.1208277.

31. Christopher E. Cornwall, Steeve Comeau, Niklas A. Kornder, Chris T. Perry, Ruben van Hooidonk, Thomas M. DeCarlo, Morgan S. Pratchett, et al., "Global Declines in Coral Reef Calcium Carbonate Production under Ocean Acidification and Warming," *Proceedings of the National Academy of Sciences* 118 (2021): e2015265118. doi: 10.1073/pnas.2015265118.

32. Kristy J. Kroeker, Rebecca L. Kordas, Ryan Crim, Iris E. Hendriks, Laura Ramajo, Gerald S. Singh, Carlos M. Duarte, et al., "Impacts of Ocean Acidification on Marine Organisms: Quantifying Sensitivities and Interaction with Warming," *Global Change Biology* 19 (2013): 1884–96. https://doi.org/10.1111/gcb.12179.

33. Kai T. Lohbeck, Ulf Riebesell, and Thorsten B. H. Reusch, "Adaptive Evolution of a Key Phytoplankton Species to Ocean Acidification," *Nature Geoscience* 5 (2012): 346–51. doi: 10.1038/ngeo1441.

34. David A. Hutchins and Feixue Fu, "Microorganisms and Ocean Global Change," *Nature Microbiology* 2 (2017): 17058. doi: 10.1038/nmicrobiol.2017.58.

35. Hutchins and Fu, "Microorganisms."

36. J. William Schopf, "Precambrian Paleobiology: Precedents, Progress, and Prospects," *Frontiers in Ecology and Evolution* 9 (2021). doi: 10.3389/fevo.2021.707072.

37. Alina Bykova, "The Changing Nature of Russia's Arctic Presence: A Case Study of Pyramiden," *The Arctic Institute* (December 9, 2019), https://www.thearcticinstitute.org/changing-nature-russia-arctic-presence-case-study-pyramiden/, accessed October 12, 2022.

38. Andrew H. Knoll, *Life on a Young Planet: The First Three Billion Years of Evolution on Earth* (Princeton, NJ: Princeton University Press, 2015).

39. M. E. Marsh, "Regulation of $CaCO_3$ Formation in Coccolithophores," *Comparative Biochemistry and Physiology Part B: Biochemistry and Molecular Biology* 136 (2003): 743–54. https://doi.org/10.1016/S1096-4959(03)00180-5.

40. Christophe Dupraz, R. Pamela Reid, Olivier Braissant, Alan W. Decho, R. Sean Norman, and Pieter T. Visscher, "Processes of Carbonate Precipitation in Modern Microbial Mats," *Earth-Science Reviews* 96 (2009): 141–62. https://doi.org/10.1016/j.earscirev.2008.10.005.

41. Dupraz et al., "Carbonate Precipitation."

42. Crisogono Vasconcelos, Judith A. McKenzie, Stefano Bernasconi, Djordje Grujic, and Albert J. Tiens, "Microbial Mediation as a Possible Mechanism for Natural Dolomite Formation at Low Temperatures," *Nature* 377 (1995): 220–22. doi: 10.1038/377220a0.

43. Susanne Douglas, "Mineralogical Footprints of Microbial Life," *American Journal of Science* 305 (2005): 503–25. doi: 10.2475/ajs.305.6-8.503.

44. E. P. Suosaari, R. P. Reid, P. E. Playford, J. S. Foster, J. F. Stolz, G. Casaburi, P. D. Hagan, et al., "New Multi-Scale Perspectives on the Stromatolites of Shark Bay, Western Australia," *Scientific Reports* 6 (2016): 20557. doi: 10.1038/srep20557.

45. Mark Feldmann and Judith A. McKenzie, "Stromatolite-Thrombolite Associations in a Modern Environment, Lee Stocking Island, Bahamas," *PALAIOS* 13 (1998): 201–12. doi: 10.2307/3515490.

46. Drake et al., "Corals Made Rocks."

47. Shanan E. Peters, Jon M. Husson, and Julia Wilcots, "The Rise and Fall of Stromatolites in Shallow Marine Environments," *Geology* 45 (2017): 487–90. doi: 10.1130/g38931.1.

48. Schopf, "Precambrian Paleobiology."

49. Martin J. Van Kranendonk, Raphael Baumgartner, Tara Djokic, Tsutomu Ota, Luke Steller, Ulf Garbe, and Eizo Nakamura, "Elements for the Origin of Life on Land: A Deep-Time Perspective from the Pilbara Craton of Western Australia," *Astrobiology* 21 (2021): 39–59. doi: 10.1089/ast.2019.2107.

50. Karl O. Stetter, "Hyperthermophiles in the History of Life," *Philosophical Transactions of the Royal Society B: Biological Sciences* 361 (2006): 1837–43. doi: 10.1098/rstb.2006.1907.

51. Andy Ridgwell and Richard E. Zeebe, "The Role of the Global Carbonate Cycle in the Regulation and Evolution of the Earth System," *Earth and Planetary Science Letters* 234 (2005): 299–315. https://doi.org/10.1016/j.epsl.2005.03.006.

52. Jonathan B. Martin, "Carbonate Minerals in the Global Carbon Cycle," *Chemical Geology* 449 (2017): 58–72. https://doi.org/10.1016/j.chemgeo.2016.11.029.

53. David Archer, *The Global Carbon Cycle* (Princeton, NJ: Princeton University Press, 2010).

54. Toby Samuels, Casey Bryce, Hanna Landenmark, Claire Marie-Loudon, Natasha Nicholson, Adam H. Stevens, and Charles Cockell, "Microbial Weathering of Minerals and Rocks in Natural Environments," in *Biogeochemical Cycles*, ed. Katerina Dontsova, Zsuzsanna Balogh-Brunstad, and Gaël Le Roux, Geophysical Monograph (Washington, DC: American Geophysical Union, 2020), pp. 59–79.

55. A. Engel Summers, A. Stern Libby, and C. Philip Bennett, "Microbial Contributions to Cave Formation; New Insights into Sulfuric Acid Speleogenesis," *Geology* 32 (2004): 369–72. doi: 10.1130/g20288.1.

56. Samuels et al., "Microbial Weathering."

57. B. Büdel, B. Weber, M. Kühl, H. Pfanz, D. Sültemeyer, and D. Wessels, "Reshaping of Sandstone Surfaces by Cryptoendolithic Cyanobacteria: Bioalkalization Causes Chemical Weathering in Arid Landscapes," *Geobiology* 2 (2004): 261–68. https://doi.org/10.1111/j.1472-4677.2004.00040.x.

58. Geoffrey Michael Gadd, "Metals, Minerals and Microbes: Geomicrobiology and Bioremediation," *Microbiology* 156 (2010): 609–43. https://doi.org/10.1099/mic.0.037143-0.

59. William Neff, "Scientific Reclamation: How the Iconic Jefferson Memorial Was Restored," *Washington Post*, October 22, 2021, https://www.washingtonpost.com/dc-md-va/interactive/2021/scientific-reclamation-how-iconic-jefferson-memorial-was-restored/, accessed February 24, 2023.

60. Andrew S. Goudie and Heather A. Viles, "Weathering and the Global Carbon Cycle: Geomorphological Perspectives," *Earth-Science Reviews* 113 (2012): 59–71. https://doi.org/10.1016/j.earscirev.2012.03.005.

61. Oleg S. Pokrovsky, Liudmila S. Shirokova, Svetlana A. Zabelina, Guntram Jordan, and Pascale Bénézeth, "Weak Impact of Microorganisms on Ca, Mg-Bearing Silicate Weathering," *npj Materials Degradation* 5 (2021): 51. doi: 10.1038/s41529-021-00199-w.

62. Goudie and Viles, "Weathering."

63. T. A. Jackson and W. D. Keller, "A Comparative Study of the Role of Lichens and 'Inorganic' Processes in the Chemical Weathering of Recent Hawaiian Lava Flows," *American Journal of Science* 269 (1970): 446–66. doi: 10.2475/ajs.269.5.446.

Chapter 7

1. Anonymous, "Methane Matters," *Nature Geoscience* 14 (2021): 875. doi: 10.1038/s41561-021-00875-1.

2. P. Forster, T. Storelvmo, K. Armour, W. Collins, J. -L. Dufresne, D. Frame, D. J. Lunt, et al., "The Earth's Energy Budget, Climate Feedbacks, and Climate Sensitivity," in *Climate Change 2021: The Physical Science Basis. Contribution of Working Group I to the Sixth Assessment Report of the Intergovernmental Panel on Climate Change*, ed. V. Masson-Delmotte, P. Zhai, A. Pirani, S. L. Connors, C. Péan , S. Berger, N. Caud, Y. Chen, L. Goldfarb, M. I. Gomis, M. Huang, K. Leitzell, E. Lonnoy, J. B. R. Matthews, T. K. Maycock, T. Waterfield, O. Yelekçi, R. Yu, and B. Zhou (Cambridge: Cambridge University Press, 2021), pp. 923–1054.

3. D. F. Ferretti, J. B. Miller, J. W. C. White, D. M. Etheridge, K. R. Lassey, D. C. Lowe, C. M. MacFarling Meure, et al., "Unexpected Changes to the Global Methane Budget over the Past 2000 Years," *Science* 309 (2005): 1714–17. doi: 10.1126/science.1115193.

4. https://gml.noaa.gov/ccgg/about.html, accessed August 8, 2022.

5. M. Saunois, A. R. Stavert, B. Poulter, P. Bousquet, J. G. Canadell, R. B. Jackson, P. A. Raymond, et al., "The Global Methane Budget 2000–2017," *Earth System Science Data* 12 (2020): 1561–623. doi: 10.5194/essd-12-1561-2020.

6. Hinrich Schaefer, Sara E. Mikaloff Fletcher, Cordelia Veidt, Keith R. Lassey, Gordon W. Brailsford, Tony M. Bromley, Edward J. Dlugokencky, et al., "A 21st-Century Shift from Fossil-Fuel to Biogenic Methane Emissions Indicated by $^{13}CH_4$," *Science* 352 (2016): 80–84. doi: 10.1126/science.aad2705.

7. Saunois et al., "Methane Budget."

8. I. S. A. Isaksen and S. B. Dalsøren, "Getting a Better Estimate of an Atmospheric Radical," *Science* 331 (2011): 38–39. doi: 10.1126/science.1199773.

9. Shushi Peng, Xin Lin, Rona L. Thompson, Yi Xi, Gang Liu, Didier Hauglustaine, Xin Lan, et al., "Wetland Emission and Atmospheric Sink Changes Explain Methane Growth in 2020," *Nature* 612 (2022): 477–82. doi: 10.1038/s41586-022-05447-w.

10. Zhen Qu, Daniel J. Jacob, Yuzhong Zhang, Lu Shen, Daniel J. Varon, Xiao Lu, Tia Scarpelli, et al., "Attribution of the 2020 Surge in Atmospheric Methane by Inverse Analysis of GOSAT Observations," *Environmental Research Letters* 17 (2022): 094003. doi: 10.1088/1748-9326/ac8754.

11. E. G. Nisbet, M. R. Manning, E. J. Dlugokencky, R. E. Fisher, D. Lowry, S. E. Michel, C. Lund Myhre, et al., "Very Strong Atmospheric Methane Growth in the 4 years 2014–2017: Implications for the Paris Agreement," *Global Biogeochemical Cycles* 33 (2019): 318–42. https://doi.org/10.1029/2018GB006009.

12. Ann R. Stavert, Marielle Saunois, Josep G. Canadell, Benjamin Poulter, Robert B. Jackson, Pierre Regnier, Ronny Lauerwald, et al., "Regional Trends and Drivers of the Global Methane Budget," *Global Change Biology* 28 (2022): 182–200. https://doi.org/10.1111/gcb.15901.

13. Saunois et al., "Methane Budget."

14. Nisbet et al., "Methane Growth," p. 327.

15. Benjamin Franklin, "Earliest Account of Marsh Gas," in *The Ingenious Dr. Franklin*, ed. Nathan G. Goodman, Selected Scientific Letters of Benjamin Franklin (University of Pennsylvania Press, 1931), pp. 145–47.

16. Thomas Paine, "The Cause of the Yellow Fever," in *The Complete Writings of Thomas Paine*, ed. Philip Sheldon Foner (New York: Citadel Press, 1945), pp. 1060–66.

17. https://www.youtube.com/watch?v=YegdEOSQotE, accessed August 22, 2022.

18. Pere Grapí, *Inspiring Air: A History of Air-Related Science* (Wilmington, DE: Vernon Press, 2019).

19. William S. Reeburgh, "Oceanic Methane Biogeochemistry," *Chemical Reviews* 107 (2007): 486–513. doi: 10.1021/cr050362v.

20. Nisbet et al., "Methane Growth."

21. Yuanlei Chen, Evan D. Sherwin, Elena S. F. Berman, Brian B. Jones, Matthew P. Gordon, Erin B. Wetherley, Eric A. Kort, et al., "Quantifying Regional Methane Emissions in the New Mexico Permian Basin with a Comprehensive Aerial Survey," *Environmental Science & Technology* 56 (2022): 4317–23. doi: 10.1021/acs.est.1c06458.

22. Sudhanshu Pandey, Ritesh Gautam, Sander Houweling, Hugo Denier van der Gon, Pankaj Sadavarte, Tobias Borsdorff, Otto Hasekamp, et al., "Satellite Observations Reveal Extreme Methane Leakage from a Natural Gas Well Blowout," *Proceedings of the National Academy of Sciences* 116 (2019): 26376–81. doi: 10.1073/pnas.1908712116.

23. Liang Feng, Paul I. Palmer, Sihong Zhu, Robert J. Parker, and Yi Liu, "Tropical Methane Emissions Explain Large Fraction of Recent Changes in Global Atmospheric Methane Growth Rate," *Nature Communications* 13 (2022): 1378. doi: 10.1038/s41467-022-28989-z.

24. Qu et al., "Atmospheric Methane."

25. Feng et al., "Tropical Methane."

26. Nisbet et al., "Methane Growth," p. 327.

27. H. Douville, K. Raghavan, J. Renwick, R. P. Allan, P. A. Arias, M. Barlow, R. Cerezo-Mota, et al., "Water Cycle Changes," in *Climate Change 2021: The Physical Science Basis. Contribution of Working Group I to the Sixth Assessment Report of the Intergovernmental Panel on Climate Change*, ed. V. Masson-Delmotte, P. Zhai, A. Pirani, S. L. Connors, C. Péan, S. Berger, N. Caud, Y. Chen, L. Goldfarb, M. I. Gomis, M. Huang, K. Leitzell, E. Lonnoy, J. B. R. Matthews, T. K. Maycock, T. Waterfield, O. Yelekçi, R. Yu, and B. Zhou (Cambridge: Cambridge University Press, 2021), pp. 1055–210.

28. Stavert et al., "Regional Trends."

29. David Bastviken, Jonathan J. Cole, Michael L. Pace, and Matthew C. Van de Bogert, "Fates of Methane from Different Lake Habitats: Connecting Whole-Lake Budgets and CH_4 Emissions," *Journal of Geophysical Research: Biogeosciences* 113 (2008): G02024. https://doi.org/10.1029/2007JG000608.

30. Jean Le Mer and Pierre Roger, "Production, Oxidation, Emission and Consumption of Methane by Soils: A Review," *European Journal of Soil Biology* 37 (2001): 25–50. https://doi.org/10.1016/S1164-5563(01)01067-6.

31. Scott D. Bridgham, Hinsby Cadillo-Quiroz, Jason K. Keller, and Qianlai Zhuang, "Methane Emissions from Wetlands: Biogeochemical, Microbial, and Modeling Perspectives from Local to Global Scales," *Global Change Biology* 19 (2013): 1325–46. https://doi.org/10.1111/gcb.12131.

32. Kimberley E. Miller, Chun-Ta Lai, Randy A. Dahlgren, and David A. Lipson, "Anaerobic Methane Oxidation in High-Arctic Alaskan Peatlands as a Significant Control on Net CH_4 Fluxes," *Soil Systems* 3 (2019). doi: 10.3390/soilsystems3010007.

33. Katrin Knittel and Antje Boetius, "Anaerobic Oxidation of Methane: Progress with an Unknown Process," *Annual Review of Microbiology* 63 (2009): 311–34. doi: 10.1146/annurev.micro.61.080706.093130.

34. Nicole Dubilier, Claudia Bergin, and Christian Lott, "Symbiotic Diversity in Marine Animals: The Art of Harnessing Chemosynthesis," *Nature Reviews Microbiology* 6 (2008): 725–40. doi: 10.1038/nrmicro1992.

35. C. K. Paull, A. J. T. Jull, L. J. Toolin, and T. Linick, "Stable Isotope Evidence for Chemosynthesis in an Abyssal Seep Community," *Nature* 317 (1985): 709–11. doi: 10.1038/317709a0.

36. Colleen M. Cavanaugh, Piet R. Levering, James S. Maki, Ralph Mitchell, and Mary E. Lidstrom, "Symbiosis of Methylotrophic Bacteria and Deep-Sea Mussels," *Nature* 325 (1987): 346–48. doi: 10.1038/325346a0.

37. Ashna A. Raghoebarsing, Alfons J. P. Smolders, Markus C. Schmid, W. Irene, C. Rijpstra, Mieke Wolters-Arts, Jan Derksen, Mike S. M. Jetten, et al., "Methanotrophic Symbionts Provide Carbon for Photosynthesis in Peat Bogs," *Nature* 436 (2005): 1153–56. doi: 10.1038/nature03802.

38. D. P. Morgavi, E. Forano, C. Martin, and C. J. Newbold, "Microbial Ecosystem and Methanogenesis in Ruminants," *Animal* 4 (2010): 1024–36. https://doi.org/10.1017/S1751731110000546.

39. Andreas Brune, "Symbiotic Digestion of Lignocellulose in Termite Guts," *Nature Reviews Microbiology* 12 (2014): 168–80. doi: 10.1038/nrmicro3182.

40. Saunois et al., "Methane Budget."
41. Ara B. Sahakian, Sam-Ryong Jee, and Mark Pimentel, "Methane and the Gastrointestinal Tract," *Digestive Diseases and Sciences* 55 (2010): 2135–43. doi: 10.1007/s10620-009-1012-0.
42. Ron Sender, Shai Fuchs, and Ron Milo, "Revised Estimates for the Number of Human and Bacteria Cells in the Body," *PLoS Biol* 14 (2016): e1002533. doi: 10.1371/journal.pbio.1002533.
43. Lenka Dohnalová, Patrick Lundgren, Jamie R. E. Carty, Nitsan Goldstein, Sebastian L. Wenski, Pakjira Nanudorn, Sirinthra Thiengmag, et al., "A Microbiome-Dependent Gut–Brain Pathway Regulates Motivation for Exercise," *Nature* 612 (2022): 739–47. doi: 10.1038/s41586-022-05525-z.
44. Annika Lenz and Lars Ojamäe, "Structures of the I-, II- and H-Methane Clathrates and the Ice–Methane Clathrate Phase Transition from Quantum-Chemical Modeling with Force-Field Thermal Corrections," *Journal of Physical Chemistry A* 115 (2011): 6169–76. doi: 10.1021/jp111328v.
45. Carolyn D. Ruppel and John D. Kessler, "The Interaction of Climate Change and Methane Hydrates," *Reviews of Geophysics* 55 (2017): 126–68. https://doi.org/10.1002/2016RG000534.
46. Vittorio Formisano, Sushil Atreya, Thérèse Encrenaz, Nikolai Ignatiev, and Marco Giuranna, "Detection of Methane in the Atmosphere of Mars," *Science* 306 (2004): 1758–61. doi: 10.1126/science.1101732.
47. E. Dendy Sloan, *Clathrate Hydrates of Natural Gases*, 2nd ed., Chemical Industries; vol. 73 (New York: Marcel Dekker, 1998).
48. Dag Vavik, "Implications of Gas Hydrates in Drilling and Completion: A Scientific Root Cause Analysis of the Deepwater Horizon Disaster," Ph.D. thesis, Norwegian University of Science and Technology, 2020. https://hdl.handle.net/11250/2651500.
49. Ruppel and Kessler, "Methane Hydrates."
50. Francesca A. McInerney and Scott L. Wing, "The Paleocene-Eocene Thermal Maximum: A Perturbation of Carbon Cycle, Climate, and Biosphere with Implications for the Future," *Annual Review of Earth and Planetary Sciences* 39 (2011): 489–516. doi: 10.1146/annurev-earth-040610-133431.
51. Bumsoo Kim and Yi Ge Zhang, "Methane Hydrate Dissociation across the Oligocene–Miocene Boundary," *Nature Geoscience* 15 (2022): 203–09. doi: 10.1038/s41561-022-00895-5.
52. Syee Weldeab,, Ralph R. Schneider, Jimin Yu, and Andrew Kylander-Clark, "Evidence for Massive Methane Hydrate Destabilization During the Penultimate Interglacial Warming," *Proceedings of the National Academy of Sciences* 119 (2022): e2201871119. doi: 10.1073/pnas.2201871119.
53. Ruppel and Kessler, "Methane Hydrates."
54. C. Berndt, T. Feseker, T. Treude, S. Krastel, V. Liebetrau, H. Niemann, V. J. Bertics, et al., "Temporal Constraints on Hydrate-Controlled Methane Seepage Off Svalbard," *Science* 343 (2014): 284–87. doi: 10.1126/science.1246298.
55. Joshua F. Dean, Jack J. Middelburg, Thomas Röckmann, Rien Aerts, Luke G. Blauw, Matthias Egger, Mike S. M. Jetten, et al., "Methane Feedbacks to the Global Climate

System in a Warmer World," *Reviews of Geophysics* 56 (2018): 207–50. https://doi.org/10.1002/2017RG000559.

56. John D. Kessler, David L. Valentine, Molly C. Redmond, Mengran Du, Eric W. Chan, Stephanie D. Mendes, Erik W. Quiroz, et al., "A Persistent Oxygen Anomaly Reveals the Fate of Spilled Methane in the Deep Gulf of Mexico," *Science* 331 (2011): 312–15. doi: 10.1126/science.1199697.

57. Yuri Shur, Daniel Fortier, M. Torre Jorgenson, Mikhail Kanevskiy, Lutz Schirrmeister, Jens Strauss, Alexander Vasiliev, et al., "Yedoma Permafrost Genesis: Over 150 Years of Mystery and Controversy," *Frontiers in Earth Science* 9 (2022). doi: 10.3389/feart.2021.757891.

58. Kimberley R. Miner, Merritt R. Turetsky, Edward Malina, Annett Bartsch, Johanna Tamminen, A. David McGuire, Andreas Fix, et al., "Permafrost Carbon Emissions in a Changing Arctic," *Nature Reviews Earth & Environment* 3 (2022): 55–67. doi: 10.1038/s43017-021-00230-3.

59. Mika Rantanen, Alexey Yu Karpechko, Antti Lipponen, Kalle Nordling, Otto Hyvärinen, Kimmo Ruosteenoja, Timo Vihma, et al., "The Arctic Has Warmed Nearly Four Times Faster Than the Globe Since 1979," *Communications Earth & Environment* 3 (2022): 168. doi: 10.1038/s43247-022-00498-3.

60. Nisbet et al., "Methane Growth."

61. Saunois et al., "Methane Budget."

62. C. D. Elder, D. R. Thompson, A. K. Thorpe, H. A. Chandanpurkar, P. J. Hanke, N. Hasson, S. R. James, et al., "Characterizing Methane Emission Hotspots from Thawing Permafrost," *Global Biogeochemical Cycles* 35 (2021): e2020GB006922. https://doi.org/10.1029/2020GB006922.

63. Merritt R. Turetsky, Benjamin W. Abbott, Miriam C. Jones, Katey Walter Anthony, David Olefeldt, Edward A. G. Schuur, Guido Grosse, et al., "Carbon Release through Abrupt Permafrost Thaw," *Nature Geoscience* 13 (2020): 138–43. doi: 10.1038/s41561-019-0526-0.

64. Christina Schädel, Martin K. F. Bader, Edward A. G. Schuur, Christina Biasi, Rosvel Bracho, Petr Čapek, Sarah De Baets, et al., "Potential Carbon Emissions Dominated by Carbon Dioxide from Thawed Permafrost Soils," *Nature Climate Change* 6 (2016): 950–53. doi: 10.1038/nclimate3054.

65. Qu et al., "Atmospheric Methane."

Chapter 8

1. Peter Nagele, Andreas Duma, Michael Kopec, Marie Anne Gebara, Alireza Parsoei, Marie Walker, Alvin Janski, et al. "Nitrous Oxide for Treatment-Resistant Major Depression: A Proof-of-Concept Trial," *Biological Psychiatry* 78 (2015): 10–18. https://doi.org/10.1016/j.biopsych.2014.11.016

2. https://www.motorbiscuit.com/engines-exposed-nitrous-oxide-is-performance-in-a-bottle/, accessed September 15, 2022.

3. https://www.nytimes.com/2022/08/31/nyregion/whipped-cream-illegal-ny.html, accessed September 22, 2022.

4. W. D. A. Smith, "A History of Nitrous Oxide and Oxygen Anaesthesia Part I: Joseph Priestley to Humphry Davy," *British Journal of Anaesthesia* 37 (1965): 790–98. https://doi.org/10.1093/bja/37.10.790.

5. Richard Holmes, *The Age of Wonder: How the Romantic Generation Discovered the Beauty and Terror of Science* (London: HarperPress, 2008).

6. Humphry Davy, *Researches, Chemical and Philosophical: Chiefly Concerning Nitrous Oxide, or Dephlogisticated Nitrous Air, and Its Respiration.* (London: J. Johnson, 1800), p. 339.

7. Davy, *Researches*, p. 487.

8. Davy, *Researches*, p. 489.

9. Sukumar P. Desai, Manisha S. Desai, and Chandrakant S. Pandav, "The Discovery of Modern Anaesthesia—Contributions of Davy, Clarke, Long, Wells and Morton," *Indian Journal of Anaesthesia* 51 (2007): 472–78.

10. E. Saikawa, R. G. Prinn, E. Dlugokencky, K. Ishijima, G. S. Dutton, B. D. Hall, R. Langenfelds, et al., "Global and Regional Emissions Estimates for N_2O," *Atmospheric Chemistry and Physics* 14 (2014): 4617–41. doi: 10.5194/acp-14-4617-2014.

11. Saikawa et al., "Emissions Estimates."

12. Adrian Schilt, Matthias Baumgartner, Thomas Blunier, Jakob Schwander, Renato Spahni, Hubertus Fischer, and Thomas F. Stocker, "Glacial–Interglacial and Millennial-Scale Variations in the Atmospheric Nitrous Oxide Concentration during the Last 800,000 Years," *Quaternary Science Reviews* 29 (2010): 182–92. https://doi.org/10.1016/j.quascirev.2009.03.011.

13. Rolf Müller, "The Impact of the Rise in Atmospheric Nitrous Oxide on Stratospheric Ozone," *Ambio* 50 (2021): 35–39. doi: 10.1007/s13280-020-01428-3.

14. M. Battle, M. Bender, T. Sowers, P. P. Tans, J. H. Butler, J. W. Elkins, J. T. Ellis, et al., "Atmospheric Gas Concentrations over the Past Century Measured in Air from Firn at the South Pole," *Nature* 383 (1996): 231–35. doi: 10.1038/383231a0.

15. Hanqin Tian, Rongting Xu, Josep G. Canadell, Rona L. Thompson, Wilfried Winiwarter, Parvadha Suntharalingam, Eric A. Davidson, et al., "A Comprehensive Quantification of Global Nitrous Oxide Sources and Sinks," *Nature* 586 (2020): 248–56. doi: 10.1038/s41586-020-2780-0.

16. E. Harris, L. Yu, Y. P. Wang, J. Mohn, S. Henne, E. Bai, M. Barthel, et al., "Warming and Redistribution of Nitrogen Inputs Drive an Increase in Terrestrial Nitrous Oxide Emission Factor," *Nature Communications* 13 (2022): 4310. doi: 10.1038/s41467-022-32001-z.

17. Tian et al., "Global Nitrous Oxide."

18. Amy Bogaard, Rebecca Fraser, Tim H. E. Heaton, Michael Wallace, Petra Vaiglova, Michael Charles, Glynis Jones, et al., "Crop Manuring and Intensive Land Management by Europe's First Farmers," *Proceedings of the National Academy of Sciences* 110 (2013): 12589–94. doi: 10.1073/pnas.1305918110.

19. Gregory T. Cushman, *Guano and the Opening of the Pacific World: A Global Ecological History* (Cambridge: Cambridge University Press, 2013).

20. Vaclav Smil, *Enriching the Earth: Fritz Haber, Carl Bosch, and the Transformation of World Food Production* (Cambridge, MA: MIT Press, 2001).

21. William Crookes, *The Wheat Problem* (London: John Murray, 1899), p. 3.

22. Collin Smith, Alfred K. Hill, and Laura Torrente-Murciano, "Current and Future Role of Haber–Bosch Ammonia in a Carbon-Free Energy Landscape," *Energy & Environmental Science* 13 (2020): 331–44. doi: 10.1039/C9EE02873K.

23. Jeremy Woods, Adrian Williams, John K. Hughes, Mairi Black, and Richard Murphy, "Energy and the Food System," *Philosophical Transactions of the Royal Society B: Biological Sciences* 365 (2010): 2991–3006. doi: 10.1098/rstb.2010.0172.

24. Smil, *Enriching the Earth*.

25. C. Lu and H. Tian, "Global Nitrogen and Phosphorus Fertilizer Use for Agriculture Production in the Past Half Century: Shifted Hot Spots and Nutrient Imbalance," *Earth System Science Data* 9 (2017): 181–92. doi: 10.5194/essd-9-181-2017.

26. Xin Zhang, Eric A. Davidson, Denise L. Mauzerall, Timothy D. Searchinger, Patrice Dumas, and Ye Shen, "Managing Nitrogen for Sustainable Development," *Nature* 528 (2015): 51–59. doi: 10.1038/nature15743.

27. Alan R. Townsend, Robert W. Howarth, Fakhri A. Bazzaz, Mary S. Booth, Cory C. Cleveland, Sharon K. Collinge, Andrew P. Dobson, et al., "Human Health Effects of a Changing Global Nitrogen Cycle," *Frontiers in Ecology and the Environment* 1 (2003): 240–46. doi: 10.1890/1540-9295(2003)001[0240:Hheoac]2.0.Co;2.

28. Bryan W. Brooks, James M. Lazorchak, Meredith D. A. Howard, Mari-Vaughn V. Johnson, Steve L. Morton, Dawn A. K. Perkins, Euan D. Reavie, et al., "Are Harmful Algal Blooms Becoming the Greatest Inland Water Quality Threat to Public Health and Aquatic Ecosystems?" *Environmental Toxicology and Chemistry* 35 (2016): 6–13. https://doi.org/10.1002/etc.3220.

29. Roslyn Wood, "Acute Animal and Human Poisonings from Cyanotoxin Exposure— A Review of the Literature," *Environment International* 91 (2016): 276–82. https://doi.org/10.1016/j.envint.2016.02.026.

30. David L. Kirchman, *Dead Zones: The Loss of Oxygen from Rivers, Lakes, Seas, and the Ocean* (New York: Oxford University Press, 2021).

31. Martin Dworkin, "Sergei Winogradsky: A Founder of Modern Microbiology and the First Microbial Ecologist," *FEMS Microbiology Reviews* 36 (2012): 364–79. 10.1111/j.1574-6976.2011.00299.x.

32. W. J. Payne, *Denitrification* (New York: Wiley, 1981).

33. Hang-Wei Hu, Deli Chen, and Ji-Zheng He, "Microbial Regulation of Terrestrial Nitrous Oxide Formation: Understanding the Biological Pathways for Prediction of Emission Rates," *FEMS Microbiology Reviews* 39 (2015): 729–49. doi: 10.1093/femsre/fuv021.

34. Tetyana Milojevic, Mihaela Albu, Denise Kölbl, Gerald Kothleitner, Robert Bruner, and Matthew L. Morgan, "Chemolithotrophy on the Noachian Martian Breccia NWA 7034 Via Experimental Microbial Biotransformation," *Communications Earth & Environment* 2 (2021): 39. doi: 10.1038/s43247-021-00105-x.

35. Dworkin, "Winogradsky."

36. Selman A. Waksman, *Sergei N. Winogradsky: His Life and Work; the Story of a Great Bacteriologist* (New Brunswick, NJ: Rutgers University Press, 1953), pp. viii–ix.

37. Dworkin, "Winogradsky."

38. Lisa Y. Stein, "The Long-Term Relationship between Microbial Metabolism and Greenhouse Gases," *Trends in Microbiology* 28 (2020): 500–11. https://doi.org/10.1016/j.tim.2020.01.006.

39. Mike S. M. Jetten, Marc Strous, Katinka T. van de Pas-Schoonen, Jos Schalk, Udo G. J. M. van Dongen, Astrid A. van de Graaf, Susanne Logemann, et al., "The Anaerobic Oxidation of Ammonium," *FEMS Microbiology Reviews* 22 (1998): 421–37. doi: 10.1111/j.1574-6976.1998.tb00379.x.

40. Xinda Lu, Anne E. Taylor, David D. Myrold, and Josh D. Neufeld, "Expanding Perspectives of Soil Nitrification to Include Ammonia-Oxidizing Archaea and Comammox Bacteria," *Soil Science Society of America Journal* 84 (2020): 287–302. https://doi.org/10.1002/saj2.20029.

41. J. Craig Venter, Karin Remington, John F. Heidelberg, Aaron L. Halpern, Doug Rusch, Jonathan A. Eisen, Dongying Wu, et al., "Environmental Genome Shotgun Sequencing of the Sargasso Sea," *Science* 304 (2004): 66–74. doi: 10.1126/science.1093857.

42. Lu et al., "Soil Nitrification."

43. Martin Könneke, Anne E. Bernhard, José R. de la Torre, Christopher B. Walker, John B. Waterbury, and David A. Stahl, "Isolation of an Autotrophic Ammonia-Oxidizing Marine Archaeon," *Nature* 437 (2005): 543–46. doi: 10.1038/nature03911.

44. Lisa Y. Stein, "Ending the Era of Haber–Bosch," *Environmental Microbiology* 25 (2023): 102–04. https://doi.org/10.1111/1462-2920.16220.

45. Lu et al., "Soil Nitrification."

46. Jonathan D. Caranto and Kyle M. Lancaster, "Nitric Oxide Is an Obligate Bacterial Nitrification Intermediate Produced by Hydroxylamine Oxidoreductase," *Proceedings of the National Academy of Sciences* 114 (2017): 8217–22. doi: 10.1073/pnas.1704504114.

47. Qixing Ji, Erik Buitenhuis, Parvadha Suntharalingam, Jorge L. Sarmiento, and Bess B. Ward, "Global Nitrous Oxide Production Determined by Oxygen Sensitivity of Nitrification and Denitrification," *Global Biogeochemical Cycles* 32 (2018): 1790–802. doi: 10.1029/2018gb005887.

48. E. R. Stuchiner and J. C. von Fischer, "Characterizing the Importance of Denitrification for N_2O Production in Soils Using Natural Abundance and Isotopic Labeling Techniques," *Journal of Geophysical Research: Biogeosciences* 127 (2022): e2021JG006555. https://doi.org/10.1029/2021JG006555.

49. Lingli Liu and Tara L. Greaver, "A Review of Nitrogen Enrichment Effects on Three Biogenic GHGs: The CO_2 Sink May Be Largely Offset by Stimulated N_2O and CH_4 Emission," *Ecology Letters* 12 (2009): 1103–17. https://doi.org/10.1111/j.1461-0248.2009.01351.x.

50. Müller, "Stratospheric Ozone."

51. Tian et al., "Global Nitrous Oxide."

52. Sara Hallin, Laurent Philippot, Frank E. Löffler, Robert A. Sanford, and Christopher M. Jones, "Genomics and Ecology of Novel N_2O-Reducing Microorganisms," *Trends in Microbiology* 26 (2018): 43–55. https://doi.org/10.1016/j.tim.2017.07.003.

53. Marcel M. M. Kuypers, Hannah K. Marchant, and Boran Kartal, "The Microbial Nitrogen-Cycling Network," *Nature Reviews Microbiology* 16 (2018): 263–76. doi: 10.1038/nrmicro.2018.9.

54. Engràcia Costa, Julio Pérez, and Jan-Ulrich Kreft, "Why Is Metabolic Labour Divided in Nitrification?," *Trends in Microbiology* 14 (2006): 213–19. https://doi.org/10.1016/j.tim.2006.03.006.

55. Srishti Vishwakarma, Xin Zhang, and Nathaniel D. Mueller, "Projecting Future Nitrogen Inputs: Are We Making the Right Assumptions?" *Environmental Research Letters* 17 (2022): 054035. doi: 10.1088/1748-9326/ac6619.

56. Tian et al., "Global Nitrous Oxide."

57. A. Landolfi, C. J. Somes, W. Koeve, L. M. Zamora, and A. Oschlies, "Oceanic Nitrogen Cycling and N_2O Flux Perturbations in the Anthropocene," *Global Biogeochemical Cycles* 31 (2017): 1236–55. https://doi.org/10.1002/2017GB005633.

58. David A. Hutchins and Douglas G. Capone, "The Marine Nitrogen Cycle: New Developments and Global Change," *Nature Reviews Microbiology* 20 (2022): 401–14. doi: 10.1038/s41579-022-00687-z.

59. Carolina Voigt, Maija E. Marushchak, Benjamin W. Abbott, Christina Biasi, Bo Elberling, Steven D. Siciliano, Oliver Sonnentag, et al., "Nitrous Oxide Emissions from Permafrost-Affected Soils," *Nature Reviews Earth & Environment* 1 (2020): 420–34. doi: 10.1038/s43017-020-0063-9.

60. Carolina Voigt, Richard E. Lamprecht, Maija E. Marushchak, Saara E. Lind, Alexander Novakovskiy, Mika Aurela, Pertti J. Martikainen, et al., "Warming of Subarctic Tundra Increases Emissions of All Three Important Greenhouse Gases—Carbon Dioxide, Methane, and Nitrous Oxide," *Global Change Biology* 23 (2017): 3121–38. https://doi.org/10.1111/gcb.13563.

61. Kimberley R. Miner, Merritt R. Turetsky, Edward Malina, Annett Bartsch, Johanna Tamminen, A. David McGuire, Andreas Fix, et al., "Permafrost Carbon Emissions in a Changing Arctic," *Nature Reviews Earth & Environment* 3 (2022): 55–67. doi: 10.1038/s43017-021-00230-3.

62. Gustaf Hugelius, Julie Loisel, Sarah Chadburn, Robert B. Jackson, Miriam Jones, Glen MacDonald, Maija Marushchak, et al., "Large Stocks of Peatland Carbon and Nitrogen Are Vulnerable to Permafrost Thaw," *Proceedings of the National Academy of Sciences* 117 (2020): 20438–46. doi: 10.1073/pnas.1916387117.

63. Fabrice Lacroix, Sönke Zaehle, Silvia Caldararu, Jörg Schaller, Peter Stimmler, David Holl, Lars Kutzbach, et al., "Mismatch of N Release from the Permafrost and Vegetative Uptake Opens Pathways of Increasing Nitrous Oxide Emissions in the High Arctic," *Global Change Biology* 28 (2022): 5973–90. https://doi.org/10.1111/gcb.16345.

64. Voigt et al., "Permafrost Review."

65. J.-Y. Lee, J. Marotzke, G. Bala, L. Cao, S. Corti, J. P. Dunne, F. Engelbrecht, E. Fischer, et al., "Future Global Climate: Scenario-Based Projections and near-Term Information," in *Climate Change 2021: The Physical Science Basis. Contribution of Working Group I to the Sixth Assessment Report of the Intergovernmental Panel on Climate Change*, ed. V. Masson-Delmotte, P. Zhai, A. Pirani, S. L. Connors, C. Péan,

S. Berger, N. Caud, Y. Chen, L. Goldfarb, M. I. Gomis, M. Huang, K. Leitzell, E. Lonnoy, J. B. R. Matthews, T. K. Maycock, T. Waterfield, O. Yelekçi, R. Yu, and B. Zhou (Cambridge: Cambridge University Press, 2021), pp. 553–672.

Chapter 9

1. https://www.usgs.gov/faqs/what-are-some-signs-climate-change, accessed February 28, 2023.
2. David A. Agar, Dimitris Athanassiadis, and Béla J. Pavelka, "The CO_2 Cutting Cost of Biogas from Humanure and Livestock Manure," *Sustainable Energy Technologies and Assessments* 53 (2022): 102381. https://doi.org/10.1016/j.seta.2022.102381.
3. https://www.eia.gov/energyexplained/biomass/landfill-gas-and-biogas.php, accessed March 19, 2023.
4. Agar et al., "Cost of Biogas."
5. Khanh Nguyen, V., Dhiraj Kumar Chaudhary, Ram Hari Dahal, N. Hoang Trinh, Jaisoo Kim, S. Woong Chang, Yongseok Hong, et al., "Review on Pretreatment Techniques to Improve Anaerobic Digestion of Sewage Sludge," *Fuel* 285 (2021): 119105. https://doi.org/10.1016/j.fuel.2020.119105.
6. Nicolae Scarlat, Jean-François Dallemand, and Fernando Fahl, "Biogas: Developments and Perspectives in Europe," *Renewable Energy* 129 (2018): 457–72. https://doi.org/10.1016/j.renene.2018.03.006.
7. https://www.eia.gov/totalenergy/, accessed February 11, 2023.
8. Scarlat et al., "Biogas in Europe."
9. https://www.eia.gov/totalenergy/, accessed February 11, 2023.
10. Shengrong Xue, Siqi Zhang, Ying Wang, Yanbo Wang, Jinghui Song, Xingang Lyu, Xiaojiao Wang, et al., "What Can We Learn from the Experience of European Countries in Biomethane Industry: Taking China as an Example?," *Renewable and Sustainable Energy Reviews* 157 (2022): 112049. https://doi.org/10.1016/j.rser.2021.112049.
11. M. C. Potter, "Electrical Effects Accompanying the Decomposition of Organic Compounds," *Proceedings of the Royal Society of London. Series B, Containing Papers of a Biological Character* 84 (1911): 260–76. doi: 10.1098/rspb.1911.0073.
12. Potter, "Electrical Effects," p. 263.
13. Bruce E. Logan, Ruggero Rossi, Ala'a Ragab, and Pascal E. Saikaly, "Electroactive Microorganisms in Bioelectrochemical Systems," *Nature Reviews Microbiology* 17 (2019): 307–19. doi: 10.1038/s41579-019-0173-x.
14. Bruce E. Logan, Bert Hamelers, René Rozendal, Uwe Schröder, Jürg Keller, Stefano Freguia, Peter Aelterman, et al., "Microbial Fuel Cells: Methodology and Technology," *Environmental Science & Technology* 40 (2006): 5181–92. doi: 10.1021/es0605016.
15. Carlo Santoro, Catia Arbizzani, Benjamin Erable, and Ioannis Ieropoulos, "Microbial Fuel Cells: From Fundamentals to Applications. A Review," *Journal of Power Sources* 356 (2017): 225–44. https://doi.org/10.1016/j.jpowsour.2017.03.109.
16. Xavier Alexis Walter, Irene Merino-Jiménez, John Greenman, and Ioannis Ieropoulos, "Pee Power® Urinal II—Urinal Scale-up with Microbial Fuel Cell Scale-Down for

Improved Lighting," *Journal of Power Sources* 392 (2018): 150–58. https://doi.org/10.1016/j.jpowsour.2018.02.047.

17. Ruggero Rossi, Andy Y. Hur, Martin A. Page, Amalia O'Brien Thomas, Joseph J. Butkiewicz, David W. Jones, Gahyun Baek, et al., "Pilot Scale Microbial Fuel Cells Using Air Cathodes for Producing Electricity While Treating Wastewater," *Water Research* 215 (2022): 118–208. https://doi.org/10.1016/j.watres.2022.118208.

18. Mariana Cardoso Chrispim, Miklas Scholz, and Marcelo Antunes Nolasco, "Biogas Recovery for Sustainable Cities: A Critical Review of Enhancement Techniques and Key Local Conditions for Implementation," *Sustainable Cities and Society* 72 (2021): 103033. https://doi.org/10.1016/j.scs.2021.103033.

19. Philip T. Pienkos and Al Darzins, "The Promise and Challenges of Microalgal-Derived Biofuels," *Biofuels, Bioproducts and Biorefining* 3 (2009): 431–40. https://doi.org/10.1002/bbb.159.

20. Arijana Bušić, Semjon Kundas, Galina Morzak, Halina Belskaya, Nenad Marđetko, Mirela Ivančić Šantek, Draženka Komes, et al., "Recent Trends in Biodiesel and Biogas Production," *Food Technology and Biotechnology* 56 (2018): 152–73. doi: 10.17113/ftb.56.02.18.5547.

21. Michael A. Borowitzka, "Energy from Microalgae: A Short History," in *Algae for Biofuels and Energy*, ed. Michael A. Borowitzka and Navid R. Moheimani (Dordrecht: Springer Netherlands, 2013), pp. 1–15.

22. https://www.greentechmedia.com/articles/read/lessons-from-the-great-algae-biof uel-bubble, accessed December 27, 2022.

23. https://corporate.exxonmobil.com/climate-solutions/advanced-biofuels, accessed December 27, 2022.

24. https://www.viridos.com/, accessed December 27, 2022.

25. Imad Ajjawi, John Verruto, Moena Aqui, Leah B. Soriaga, Jennifer Coppersmith, Kathleen Kwok, Luke Peach, et al., "Lipid Production in *Nannochloropsis gaditana* Is Doubled by Decreasing Expression of a Single Transcriptional Regulator," *Nature Biotechnology* 35 (2017): 647–52. 10.1038/nbt.3865.

26. Amy Westervelt, "Big Oil Firms Touted Algae as Climate Solution. Now All Have Pulled Funding," *The Guardian*, March 17, 2023, https://www.theguardian.com/environment/2023/mar/17/big-oil-algae-biofuel-funding-cut-exxonmobil, accessed March 28, 2023.

27. A. Bušić, N. Marđetko, S. Kundas, G. Morzak, H. Belskaya, M. Ivančić Šantek, D. Komes, et al., "Bioethanol Production from Renewable Raw Materials and Its Separation and Purification: A Review," *Food Technology and Biotechnology* 56 (2018): 289–311. doi: 10.17113/ftb.56.03.18.5546.

28. Azhar Mohd, Siti Hajar, Rahmath Abdulla, Siti Azmah Jambo, Hartinie Marbawi, Jualang Azlan Gansau, Ainol Azifa Mohd Faik, and Kenneth Francis Rodrigues, "Yeasts in Sustainable Bioethanol Production: A Review," *Biochemistry and Biophysics Reports* 10 (2017): 52–61. https://doi.org/10.1016/j.bbrep.2017.03.003.

29. Magdalena Broda, Daniel J. Yelle, and Katarzyna Serwańska, "Bioethanol Production from Lignocellulosic Biomass—Challenges and Solutions," *Molecules* 27 (2022): 8717. doi: 10.3390/molecules27248717.

30. https://www.fas.usda.gov/data/philippines-biofuels-annual-7, accessed January 2, 2023.

31. Bušić et al., "Bioethanol."

32. https://www.iea.org/, accessed December 29, 2022.

33. Bušić et al., "Bioethanol."

34. C. Manochio, B. R. Andrade, R. P. Rodriguez, and B. S. Moraes, "Ethanol from Biomass: A Comparative Overview," *Renewable and Sustainable Energy Reviews* 80 (2017): 743–55. https://doi.org/10.1016/j.rser.2017.05.063.

35. Tyler J. Lark, Nathan P. Hendricks, Aaron Smith, Nicholas Pates, Seth A. Spawn-Lee, Matthew Bougie, Eric G. Booth, et al., "Environmental Outcomes of the US Renewable Fuel Standard," *Proceedings of the National Academy of Sciences* 119 (2022): e2101084119. doi: 10.1073/pnas.2101084119.

36. Kamran Malik, Priyanka Sharma, Yulu Yang, Peng Zhang, Lihong Zhang, Xiaohong Xing, Jianwei Yue, et al., "Lignocellulosic Biomass for Bioethanol: Insight into the Advanced Pretreatment and Fermentation Approaches," *Industrial Crops and Products* 188 (2022): 115569. https://doi.org/10.1016/j.indcrop.2022.115569.

37. https://www.eia.gov/, accessed February 11, 2023.

38. M. Saunois, A. R. Stavert, B. Poulter, P. Bousquet, J. G. Canadell, R. B. Jackson, P. A. Raymond, et al., "The Global Methane Budget 2000–2017," *Earth System Science Data* 12 (2020): 1561–623. doi: 10.5194/essd-12-1561-2020.

39. J. Poore and T. Nemecek, "Reducing Food's Environmental Impacts through Producers and Consumers," *Science* 360 (2018): 987–92. doi: 10.1126/science.aaq0216.

40. Lei Zhang, Hanqin Tian, Hao Shi, Shufen Pan, Jinfeng Chang, Shree R. S. Dangal, Xiaoyu Qin, et al., "A 130-Year Global Inventory of Methane Emissions from Livestock: Trends, Patterns, and Drivers," *Global Change Biology* 28 (2022): 5142–58. https://doi.org/10.1111/gcb.16280.

41. Percentages given in Zhang et al., "Global Inventory" were corrected for methane emissions from rice reported by Saunois et al., "Methane Budget."

42. Zhang et al., "Global Inventory."

43. Laura Reiley, "Moooove Over: How Single-Celled Yeasts Are Doing the Work of 1,500-Pound Cows," *Washington Post*, March 12, 2023, https://www.washingtonpost.com/business/2023/03/12/precision-cultivated-dairy/, accessed March 12, 2023.

44. Vera Sokolov, Andrew VanderZaag, Jemaneh Habtewold, Kari Dunfield, James T. Tambong, Claudia Wagner-Riddle, Jason J. Venkiteswaran, et al., "Acidification of Residual Manure in Liquid Dairy Manure Storages and Its Effect on Greenhouse Gas Emissions," *Frontiers in Sustainable Food Systems* 4 (2020): 568648. doi: 10.3389/fsufs.2020.568648.

45. Brendan P. Harrison, Si Gao, Melinda Gonzales, Touyee Thao, Elena Bischak, Teamrat Afewerki Ghezzehei, Asmeret Asefaw Berhe, et al., "Dairy Manure Co-Composting with Wood Biochar Plays a Critical Role in Meeting Global Methane Goals," *Environmental Science & Technology* 56 (2022): 10987–96. doi: 10.1021/acs.est.2c03467.

46. J. B. Russell and J. L. Rychlik, "Factors That Alter Rumen Microbial Ecology," *Science* 292 (2001): 1119–22. doi: 10.1126/science.1058830.

47. Guanghui Yu, Karen A. Beauchemin, and Ruilan Dong, "A Review of 3-Nitrooxypropanol for Enteric Methane Mitigation from Ruminant Livestock," *Animals* 11 (2021): 3540. doi: 10.3390/ani11123540.

48. Somini Sengupta, "Seaweed Is Having Its Moment in the Sun," *New York Times*, March 15, 2023, https://www.nytimes.com/interactive/2023/03/15/climate/seaweed-plastic-climate-change.html?campaign_id=9&emc=edit_nn_20230316&instance_id=87830&nl=the-morning&regi_id=50790297&segment_id=127903&te=1&user_id=e3065bae56ca266ff8ff60074fc4da2d, accessed March 19, 2023.

49. Jinyang Wang, Philippe Ciais, Pete Smith, Xiaoyuan Yan, Yakov Kuzyakov, Shuwei Liu, Tingting Li, et al., "The Role of Rice Cultivation in Changes in Atmospheric Methane Concentration and the Global Methane Pledge," *Global Change Biology* 29 (2023): 2776–89. https://doi.org/10.1111/gcb.16631.

50. Youzhi Feng, Yanping Xu, Yongchang Yu, Zubin Xie, and Xiangui Lin, "Mechanisms of Biochar Decreasing Methane Emission from Chinese Paddy Soils," *Soil Biology and Biochemistry* 46 (2012): 80–88. https://doi.org/10.1016/j.soilbio.2011.11.016.

51. Xingguo Han, Xue Sun, Cheng Wang, Mengxiong Wu, Da Dong, Ting Zhong, Janice E. Thies, et al., "Mitigating Methane Emission from Paddy Soil with Rice-Straw Biochar Amendment under Projected Climate Change," *Scientific Reports* 6 (2016): 24731. doi: 10.1038/srep24731.

52. Charlotte Decock, "Mitigating Nitrous Oxide Emissions from Corn Cropping Systems in the Midwestern US: Potential and Data Gaps," *Environmental Science & Technology* 48 (2014): 4247–56. doi: 10.1021/es4055324.

53. Rodney T. Venterea, Jeffrey A. Coulter, and Michael S. Dolan, "Evaluation of Intensive "4R" Strategies for Decreasing Nitrous Oxide Emissions and Nitrogen Surplus in Rainfed Corn," *Journal of Environmental Quality* 45 (2016): 1186–95. https://doi.org/10.2134/jeq2016.01.0024.

54. Isaac M. Klimasmith and Angela D. Kent, "Micromanaging the Nitrogen Cycle in Agroecosystems," *Trends in Microbiology* 30 (2022): 1045–55. https://doi.org/10.1016/j.tim.2022.04.006.

55. Klimasmith and Kent, "Nitrogen Cycle."

56. Manabu Itakura, Yoshitaka Uchida, Hiroko Akiyama, Yuko Takada Hoshino, Yumi Shimomura, Sho Morimoto, Kanako Tago, et al., "Mitigation of Nitrous Oxide Emissions from Soils by *Bradyrhizobium japonicum* Inoculation," *Nature Climate Change* 3 (2013): 208–12. doi: 10.1038/nclimate1734.

57. Amy Wen, Keira L. Havens, Sarah E. Bloch, Neal Shah, Douglas A. Higgins, Austin G. Davis-Richardson, Judee Sharon, et al., "Enabling Biological Nitrogen Fixation for Cereal Crops in Fertilized Fields," *ACS Synthetic Biology* 10 (2021): 3264–77. doi: 10.1021/acssynbio.1c00049.

58. https://www.pivotbio.com/, accessed January 9, 2023.

59. Michael R. Silverstein, Daniel Segrè, and Jennifer M. Bhatnagar, "Environmental Microbiome Engineering for the Mitigation of Climate Change," *Global Change Biology* 29 (2013): 2050–66. https://doi.org/10.1111/gcb.16609.

60. Mariana Sanches Santos, Marco Antonio Nogueira, and Mariangela Hungria, "Microbial Inoculants: Reviewing the Past, Discussing the Present and Previewing an

Outstanding Future for the Use of Beneficial Bacteria in Agriculture," *AMB Express* 9 (2019): 205. doi: 10.1186/s13568-019-0932-0.

61. https://biosolutions.novozymes.com/en/products/forages/nitragin-gold-alfalfa-and-sweet-clover, accessed March 1, 2023.

62. Jos M. Raaijmakers and E. Toby Kiers, "Rewilding Plant Microbiomes," *Science* 378 (2022): 599–600. doi: 10.1126/science.abn6350.

63. Silverstein et al., "Environmental Microbiome Engineering."

64. Nathaniel C. Lawrence, Carlos G. Tenesaca, Andy VanLoocke, and Steven J. Hall, "Nitrous Oxide Emissions from Agricultural Soils Challenge Climate Sustainability in the US Corn Belt," *Proceedings of the National Academy of Sciences* 118 (2021): e2112108118. doi: 10.1073/pnas.2112108118.

65. J.-Y. Lee, J. Marotzke, G. Bala, L. Cao, S. Corti J. P. Dunne, F. Engelbrecht, E. Fischer, et al., "Future Global Climate: Scenario-Based Projections and Near-Term Information," in *Climate Change 2021: The Physical Science Basis. Contribution of Working Group I to the Sixth Assessment Report of the Intergovernmental Panel on Climate Change*, ed. V. Masson-Delmotte, P. Zhai, A. Pirani, S. L. Connors, C. Péan, S. Berger, N. Caud, Y. Chen, L. Goldfarb, M. I. Gomis, M. Huang, K. Leitzell, E. Lonnoy, J. B. R. Matthews, T. K. Maycock, T. Waterfield, O. Yelekçi, R. Yu, and B. Zhou (Cambridge: Cambridge University Press, 2021), pp. 553–672.

66. G. Philip Robertson, Stephen K. Hamilton, Keith Paustian, and Pete Smith, "Land-Based Climate Solutions for the United States," *Global Change Biology* 28 (2022): 4912–19. https://doi.org/10.1111/gcb.16267.

67. P. Friedlingstein, M. O'Sullivan, M. W. Jones, R. M. Andrew, L. Gregor, J. Hauck, C. Le Quéré, et al. "Global Carbon Budget 2022." *Earth System Science Data* 14 (2022): 4811–900. doi: 10.5194/essd-14-4811-2022.

68. Cesare Marchetti, "On Geoengineering and the CO_2 Problem," *Climatic Change* 1 (1977): 59–68. doi: 10.1007/BF00162777, p. 61.

69. Ehsan Daneshvar, Rebecca J. Wicker, Pau-Loke Show, and Amit Bhatnagar, "Biologically-Mediated Carbon Capture and Utilization by Microalgae Towards Sustainable CO_2 Biofixation and Biomass Valorization—A Review," *Chemical Engineering Journal* 427 (2022): 130884. https://doi.org/10.1016/j.cej.2021.130884.

70. Ambreen Aslam, Skye R. Thomas-Hall, Tahira Aziz Mughal, and Peer M. Schenk, "Selection and Adaptation of Microalgae to Growth in 100% Unfiltered Coal-Fired Flue Gas," *Bioresource Technology* 233 (2017): 271–83. https://doi.org/10.1016/j.biortech.2017.02.111.

71. https://www.globalccsinstitute.com/, accessed January 13, 2023.

72. An emission of 37 billion metric tons of carbon dioxide is equivalent to about 10 petagrams of carbon, as discussed in Chapter 2.

73. Brad Plumer, "A Huge City Polluter? Buildings. Here's a Surprising Fix," *New York Times*, March 10, 2023, https://www.nytimes.com/interactive/2023/03/10/climate/buildings-carbon-dioxide-emissions-climate.html, accessed March 11, 2023.

74. https://www.algiecel.com/, accessed January 14, 2023.

75. Oxford English Dictionary, https://www.oed.com/, accessed January 14, 2023.

76. Kenneth Roy, "The Solar Shield Concept: Current Status and Future Possibilities," *Acta Astronautica* 197 (2022): 368–74. https://doi.org/10.1016/j.actaastro.2022.02.022.

77. T. M. Lenton and N. E. Vaughan, "The Radiative Forcing Potential of Different Climate Geoengineering Options," *Atmospheric Chemistry and Physics* 9 (2009): 5539–61. doi: 10.5194/acp-9-5539-2009.

78. https://www.nature.org/en-us/get-involved/how-to-help/plant-a-billion/, accessed January 17, 2023.

79. Wayne S. Walker, Seth R. Gorelik, Susan C. Cook-Patton, Alessandro Baccini, Mary K. Farina, Kylen K. Solvik, Peter W. Ellis, et al., "The Global Potential for Increased Storage of Carbon on Land," *Proceedings of the National Academy of Sciences* 119 (2022): e2111312119. doi: 10.1073/pnas.2111312119.

80. Mark A. Anthony, Thomas W. Crowther, Sietse van der Linde, Laura M. Suz, Martin I. Bidartondo, Filipa Cox, Marcus Schaub, et al., "Forest Tree Growth Is Linked to Mycorrhizal Fungal Composition and Function across Europe," *The ISME Journal* 16 (2022): 1327–36. doi: 10.1038/s41396-021-01159-7.

81. Walker et al., "Global Potential."

82. Dominic Woolf, James E. Amonette, F. Alayne Street-Perrott, Johannes Lehmann, and Stephen Joseph, "Sustainable Biochar to Mitigate Global Climate Change," *Nature Communications* 1 (2010): 56. doi: 10.1038/ncomms1053.

83. Morgan J. Schmidt, Anne Rapp Py-Daniel, Claide de Paula Moraes, Raoni B. M. Valle, Caroline F. Caromano, Wenceslau G. Texeira, Carlos A. Barbosa, et al., "Dark Earths and the Human Built Landscape in Amazonia: A Widespread Pattern of Anthrosol Formation," *Journal of Archaeological Science* 42 (2014): 152–65. https://doi.org/10.1016/j.jas.2013.11.002.

84. Humberto Blanco-Canqui, David A. Laird, Emily A. Heaton, Samuel Rathke, and Bharat Sharma Acharya, "Soil Carbon Increased by Twice the Amount of Biochar Carbon Applied after 6 Years: Field Evidence of Negative Priming," *GCB Bioenergy* 12 (2020): 240–51. https://doi.org/10.1111/gcbb.12665.

85. C. Werner, W. Lucht, D. Gerten, and C. Kammann, "Potential of Land-Neutral Negative Emissions through Biochar Sequestration," *Earth's Future* 10 (2022): e2021EF002583. https://doi.org/10.1029/2021EF002583.

86. Yuanyuan Huang, Phillipe Ciais, Yiqi Luo, Dan Zhu, Yingping Wang, Chunjing Qiu, Daniel S. Goll, et al., "Tradeoff of CO_2 and CH_4 Emissions from Global Peatlands under Water-Table Drawdown," *Nature Climate Change* 11 (2021): 618–22. doi: 10.1038/s41558-021-01059-w.

87. Junyu Zou, Alan D. Ziegler, Deliang Chen, Gavin McNicol, Philippe Ciais, Xin Jiang, Chunmiao Zheng, et al., "Rewetting Global Wetlands Effectively Reduces Major Greenhouse Gas Emissions," *Nature Geoscience* 15 (2022): 627–32. doi: 10.1038/s41561-022-00989-0.

88. Virginia Gewin, "How Peat Could Protect the Planet," *Nature* 578 (2020): 204–8. https://doi.org/10.1038/d41586-020-00355-3.

89. Ranran Zhou, Yuan Liu, Jennifer A. J. Dungait, Amit Kumar, Jinsong Wang, Lisa K. Tiemann, Fusuo Zhang, et al., "Microbial Necromass in Cropland Soils: A Global

Meta-Analysis of Management Effects," *Global Change Biology* 29 (2023): 1998–2014. https://doi.org/10.1111/gcb.16613.

90. Yongfei Bai and M. Francesca Cotrufo, "Grassland Soil Carbon Sequestration: Current Understanding, Challenges, and Solutions," *Science* 377 (2022): 603–8. doi: 10.1126/science.abo2380.

91. https://4p1000.org, accessed January 18, 2023.

92. National Academies of Sciences, Engineering, and Medicine, *A Research Strategy for Ocean-Based Carbon Dioxide Removal and Sequestration* (Washington, DC: The National Academies Press, 2022).

93. National Academies of Sciences, *Research Strategy.*

94. Jeff Tollefson, "Ocean-Fertilization Project Off Canada Sparks Furore," *Nature* 490 (2012): 458–59. doi: 10.1038/490458a.

95. National Academies of Sciences, *Research Strategy.*

96. Tollefson, "Ocean Fertilization."

97. National Academies of Sciences, *Research Strategy.*

Glossary

abiotic Physical or chemical, not a living organism.

aerobic A metabolism that depends on oxygen.

algae Organisms that include microbes and seaweeds capable of carrying out photosynthesis like plants but without leaves, stems, true roots, or a vascular system. The term is often used to mean microalgae. The singular is "alga."

anaerobic A metabolism that does not depend on oxygen.

anoxic A habitat without oxygen gas.

anthropogenic Due to human activity. Most of climate change now occurring is anthropogenic, due to the use of fossil fuels and other human actions.

archaea Microbes that do not have internal organelles such as a nucleus to house their genome. They superficially resemble bacteria but are a separate domain of life. An example is methanogens. The singular is "archaeon."

bacteria Microbes that do not have internal organelles such as a nucleus to house their genome. They superficially resemble archaea but are a separate domain of life. Some examples include the pathogens that cause cholera (*Vibrio cholerae*) and food poisoning (*Salmonella*). The singular is "bacterium."

biological pump A process by which carbon dioxide consumed by phytoplankton at the surface is transported to deep waters and sediments.

biomass The quantity or amount of material in living organisms.

biosphere The parts of Earth occupied by living organisms, the composite of all ecosystems.

bloom A buildup of algae or phytoplankton over a short period as a result of vigorous growth and lack of predation. Some algal blooms can be toxic, but most are crucial for supporting aquatic food webs.

carbon cycle The conversions between inorganic carbon, such as carbon dioxide, and organic carbon in and between the atmosphere, land, and the ocean.

carbon dioxide fixation The conversion of the gas carbon dioxide to nongaseous organic chemicals. The reaction occurs during photosynthesis.

chlorophyll A pigment used by plants, algae, and cyanobacteria to carry out photosynthesis. It is used by ecologists and climate-change scientists as an index of land plant or phytoplankton biomass.

climate change Long-term variation in average temperature, precipitation, and other aspects of weather.

cyanobacteria Microbes that carry out photosynthesis like plants and algae but are bacteria. An old name for cyanobacteria is "blue-green algae."

diffusion The movement of chemicals, gases, or heat from high to low levels in air or water.

electron acceptor A chemical that takes on electrons during the oxidation of another chemical such as a sugar. Common electron acceptors are oxygen gas, sulfate, and nitrate.

electron donor A chemical like sugar that gives up its electrons to another chemical like oxygen gas. In addition to organic chemicals, inorganic ones like ammonia and sulfide can also be electron donors.

eukaryote An organism with organelles such as a nucleus to house its genome. Examples include animals, plants, fungi, and protozoa. Bacteria and archaea are not eukaryotes.

fixed nitrogen A nitrogen chemical that is not a gas like nitrogen gas or nitrous oxide. Examples include nitrate and protein.

flux The movement of a chemical from one pool or reservoir to another.

food chain Linked predator–prey interactions in which one organism eats another.

fossil fuels Energy sources such as petroleum, coal, and natural gas, formed millions of years ago.

fungus Eukaryotic organisms that feed on dead organic matter. Some examples include yeasts, molds, and mushrooms. The plural is "fungi."

geoengineering The deliberate, large-scale manipulation of soils, oceans, or the atmosphere to combat climate change.

global warming The increase in Earth's average temperature over a long time. It is one manifestation of climate change.

greenhouse effect The trapping of heat from sunlight near Earth's surface.

greenhouse gas A gas capable of trapping heat from sunlight near Earth's surface. Three examples are carbon dioxide, methane, and nitrous oxide.

heterotroph An organism that cannot make its own organic chemicals needed to make cellular components and must consume organics made by autotrophs. Examples include animals, fungi, and many kinds of bacteria.

hypoxia A condition or habitat with low oxygen gas.

inorganic chemical Not an organic chemical. Bicarbonate and carbonate are two forms of inorganic carbon.

IPCC Abbreviation for Intergovernmental Panel on Climate Change, the United Nations group that assesses climate change science.

isotope One of two or more forms of an element that have the same number of protons but differ in the number of neutrons and thus molecular weight. Carbon isotopes include ^{12}C (the most abundant one), ^{13}C (stable and nonradioactive), and ^{14}C (unstable and radioactive).

labile A chemical that is easy to use, degrade, or decompose by microbes.

macroalgae Seaweeds capable of carrying out photosynthesis like plants but are without leaves, stems, true roots, or a vascular system. An example is kelp.

methanogenesis The making of methane, which is mostly by archaea.

methanotrophy The degradation of methane.

microalgae A type of eukaryotic microbe that carries out photosynthesis like plants.

microbe Any organism that requires a microscope to be seen. Examples include bacteria, microalgae, some fungi, and protists.

mycorrhiza A type of fungus living symbiotically with plant roots.

necromass Organic matter that comes from microbes.

negative feedback A change that leads to a reduction in global warming or another manifestation of climate change. As an example, global warming may cause microbes to produce more cloud condensation nuclei and more clouds that block sunlight and lower temperatures.

organic chemical A molecule or compound that contains carbon bound to another carbon atom or to a hydrogen atom. Examples include sugar, cellulose, methane, and petroleum. Carbon dioxide is not an organic chemical.

organic matter The sum of all organic chemicals in a sample.

oxic A habitat with oxygen.

oxidation The transformation of a chemical to another by the loss of electrons after contact with oxygen or another electron acceptor. The complete oxidation or "burning" of an organic chemical yields carbon dioxide.

photosynthesis The synthesis of organic chemicals using carbon dioxide and light energy, carried out by cyanobacteria, algae, and plants.

phytoplankton Microalgae and cyanobacteria free-floating in an aquatic habitat.

positive feedback A change that leads to an additional increase in global warming or another manifestation of climate change. For example, warming causes more carbon dioxide to be released, which in turn leads to more warming.

primary production The first step in a food chain where organisms use sunlight energy and carbon dioxide to grow and reproduce. The organisms carrying out primary production include plants, algae, and cyanobacteria.

prokaryote A microbe without a nucleus to house its genome. Bacteria and archaea are prokaryotes, whereas eukaryotes are not.

protist A single-cell eukaryote, such as microalgae and protozoa.

protozoa A protist that carries out an animal-like metabolism. Examples include ciliates, rotifers, and some flagellates. The singular is "protozoan."

reduction The transformation of a chemical to another by gaining electrons from an electron donor. An example is the transformation of carbon dioxide to an organic chemical.

refractory organic matter Organic chemicals that are hard to degrade or decompose.

respiration The process by which organisms gain energy by combining a chemical like sugar with an electron acceptor, like oxygen, releasing carbon dioxide and water. In aerobic respiration, the electron acceptor is oxygen, whereas in anaerobic respiration, it is another chemical such as nitrate or sulfate.

ruminant An animal with a rumen, a stomach-like sac where plant material is digested. Examples include cows, sheep, and deer.

sedimentary rock A rock that forms from the accumulation of minerals or organic matter from organisms. Limestone is one example.

sequestration The storage of carbon away from the atmosphere.

sink An ecosystem or organism that takes in more of a chemical than it releases. Land plants and the oceans are large sinks for carbon dioxide.

stock The amount or quantity of something, such as the amount of carbon in soils.

symbiont An organism that lives physically close to another organism in a symbiotic relationship. According to the definition of symbiosis used in this book, both partners benefit from the relationship.

virus A biological entity that must infect a cell to reproduce. Outside of a cellular host, it is inert without any metabolism. Although viruses can be seen only with a microscope, they are not microbes.

weathering The breakdown of rocks when in contact with water, gases, or microbes.

zooplankton Small organisms with an animal-like metabolism that cannot swim effectively against currents and are free-floating in the water column. Many are microscopic, but some can be seen with the naked eye.

Selected Bibliography

Chapter 1

Falkowski, Paul G. *Life's Engines: How Microbes Made Earth Habitable*. Science Essentials. Princeton, NJ: Princeton University Press, 2015.

Friedlingstein, P., M. W. Jones, M. O'Sullivan, R. M. Andrew, D. C. E. Bakker, J. Hauck, C. Le Quéré, et al. "Global Carbon Budget 2021." *Earth System Science Data* 14 (2022): 1917–2005. doi: 10.5194/essd-2021-386.

Keeling, Charles D. "Rewards and Penalties of Monitoring the Earth." *Annual Review of Energy and the Environment* 23 (1998): 25–82. doi: 10.1146/annurev.energy.23.1.25.

Chapter 2

Bar-On, Yinon M., Rob Phillips, and Ron Milo. "The Biomass Distribution on Earth." *Proceedings of the National Academy of Sciences* 115 (2018): 6506–11. doi: 10.1073/pnas.1711842115.

Field, C. B., M. J. Behrenfeld, J. T. Randerson, and P. Falkowski. "Primary Production of the Biosphere: Integrating Terrestrial and Oceanic Components." *Science* 281 (1998): 237–40. doi: 10.1126/science.281.5374.237.

Jian, J., R. Vargas, K. Anderson-Teixeira, E. Stell, V. Herrmann, M. Horn, N. Kholod, et al. "A Restructured and Updated Global Soil Respiration Database (Srdb-V5)." *Earth System Science Data* 13 (2021): 255–67. doi: 10.5194/essd-13-255-2021.

Robinson, Carol. "Microbial Respiration, the Engine of Ocean Deoxygenation." *Frontiers in Marine Science* 5 (2019). doi: 10.3389/fmars.2018.00533.

Chapter 3

Bond-Lamberty, Ben, Vanessa L. Bailey, Min Chen, Christopher M. Gough, and Rodrigo Vargas. "Globally Rising Soil Heterotrophic Respiration over Recent Decades." *Nature* 560 (2018): 80–83. doi: 10.1038/s41586-018-0358-x.

Camenzind, Tessa, Kyle Mason-Jones, India Mansour, Matthias C. Rillig, and Johannes Lehmann. "Formation of Necromass-Derived Soil Organic Carbon Determined by Microbial Death Pathways." *Nature Geoscience* 16 (2023): 115–22. doi: 10.1038/s41561-022-01100-3.

Schmidt, Michael W. I., Margaret S. Torn, Samuel Abiven, Thorsten Dittmar, Georg Guggenberger, Ivan A. Janssens, Markus Kleber, et al. "Persistence of Soil Organic Matter as an Ecosystem Property." *Nature* 478 (2011): 49–56. doi: 10.1038/nature10386.

Schuur, Edward A. G., Rosvel Bracho, Gerardo Celis, E. Fay Belshe, Chris Ebert, Justin Ledman, Marguerite Mauritz, et al. "Tundra Underlain by Thawing Permafrost

Persistently Emits Carbon to the Atmosphere over 15 Years of Measurements." *Journal of Geophysical Research: Biogeosciences* 126 (2021): e2020JG006044. https://doi.org/10.1029/2020JG006044.

Sokol, Noah W., Eric Slessarev, Gianna L. Marschmann, Alexa Nicolas, Steven J. Blazewicz, Eoin L. Brodie, Mary K. Firestone, et al. "Life and Death in the Soil Microbiome: How Ecological Processes Influence Biogeochemistry." *Nature Reviews Microbiology* 20 (2022): 415–30. doi: 10.1038/s41579-022-00695-z.

Chapter 4

Dittmar, Thorsten, Sinikka T. Lennartz, Hagen Buck-Wiese, Dennis A. Hansell, Chiara Santinelli, Chiara Vanni, Bernd Blasius, et al. "Enigmatic Persistence of Dissolved Organic Matter in the Ocean." *Nature Reviews Earth & Environment* 2 (2021): 570–83. doi: 10.1038/s43017-021-00183-7.

Hutchins, David A., and Feixue Fu. "Microorganisms and Ocean Global Change." *Nature Microbiology* 2 (2017): 17058. doi: 10.1038/nmicrobiol.2017.58.

Moran, Mary Ann, Frank X. Ferrer-González, He Fu, Brent Nowinski, Malin Olofsson, McKenzie A. Powers, Jeremy E. Schreier, et al. "The Ocean's Labile DOC Supply Chain." *Limnology and Oceanography* 67 (2022): 1007–21. https://doi.org/10.1002/lno.12053.

Nowicki, Michael, Tim DeVries, and David A. Siegel. "Quantifying the Carbon Export and Sequestration Pathways of the Ocean's Biological Carbon Pump." *Global Biogeochemical Cycles* 36 (2022): e2021GB007083. https://doi.org/10.1029/2021GB007083.

Schmidtko, Sunke, Lothar Stramma, and Martin Visbeck. "Decline in Global Oceanic Oxygen Content During the Past Five Decades." *Nature* 542 (2017): 335–39. doi: 10.1038/nature21399.

Wilson, Jamie D., Oliver Andrews, Anna Katavouta, Francisco de Melo Viríssimo, Ros M. Death, Markus Adloff, Chelsey A. Baker, et al. "The Biological Carbon Pump in CMIP6 Models: 21st Century Trends and Uncertainties." *Proceedings of the National Academy of Sciences* 119 (2022): e2204369119. doi:10.1073/pnas.2204369119.

Chapter 5

Charlson, R. J., J. E. Lovelock, M. O. Andreae, and S. G. Warren. "Oceanic Phytoplankton, Atmospheric Sulfur, Cloud Albedo and Climate." *Nature* 326 (1987): 655–61. https://doi.org/10.1038/326655a0.

Moran, Mary Ann, Chris R. Reisch, Ronald P. Kiene, and William B. Whitman. "Genomic Insights into Bacterial DMSP Transformations." *Annual Review of Marine Science* 4 (2012): 523–42. doi: 10.1146/annurev-marine-120710-100827.

Sellegri, Karine, Alessia Nicosia, Evelyn Freney, Julia Uitz, Melilotus Thyssen, Gérald Grégori, Anja Engel, et al. "Surface Ocean Microbiota Determine Cloud Precursors." *Scientific Reports* 11 (2021): 281. doi: 10.1038/s41598-020-78097-5.

Tyrrell, Toby. *On Gaia: A Critical Investigation of the Relationship between Life and Earth.* Princeton, NJ: Princeton University Press, 2013.

Wang, Shanlin, Mathew Maltrud, Scott Elliott, Philip Cameron-Smith, and Alexandra Jonko. "Influence of Dimethyl Sulfide on the Carbon Cycle and Biological Production." *Biogeochemistry* 138 (2018): 49–68. doi: 10.1007/s10533-018-0430-5.

Chapter 6

Cornwall, Christopher E., Steeve Comeau, Niklas A. Kornder, Chris T. Perry, Ruben van Hooidonk, Thomas M. DeCarlo, Morgan S. Pratchett, et al. "Global Declines in Coral Reef Calcium Carbonate Production under Ocean Acidification and Warming." *Proceedings of the National Academy of Sciences* 118 (2021): e2015265118. doi: 10.1073/pnas.2015265118.

Doney, Scott C., D. Shallin Busch, Sarah R. Cooley, and Kristy J. Kroeker. "The Impacts of Ocean Acidification on Marine Ecosystems and Reliant Human Communities." *Annual Review of Environment and Resources* 45 (2020): 83–112. doi: 10.1146/annurev-environ-012320-083019.

Drake, Jeana L., Tali Mass, Jarosław Stolarski, Stanislas Von Euw, Bas van de Schootbrugge, and Paul G. Falkowski. "How Corals Made Rocks through the Ages." *Global Change Biology* 26 (2020): 31–53. https://doi.org/10.1111/gcb.14912.

Hopkins, Jason, Stephanie A. Henson, Stuart C. Painter, Toby Tyrrell, and Alex J. Poulton. "Phenological Characteristics of Global Coccolithophore Blooms." *Global Biogeochemical Cycles* 29 (2015): 239–53. https://doi.org/10.1002/2014GB004919.

Iglesias-Rodríguez, M. Débora, Christopher W. Brown, Scott C. Doney, Joan Kleypas, Dorota Kolber, Zbigniew Kolber, Paul K. Hayes, et al. "Representing Key Phytoplankton Functional Groups in Ocean Carbon Cycle Models: Coccolithophorids." *Global Biogeochemical Cycles* 16 (2002): 47-1–47-20. https://doi.org/10.1029/2001GB001454.

Samuels, Toby, Casey Bryce, Hanna Landenmark, Claire Marie-Loudon, Natasha Nicholson, Adam H. Stevens, and Charles Cockell. "Microbial Weathering of Minerals and Rocks in Natural Environments." In *Biogeochemical Cycles*, edited by Katerina Dontsova, Zsuzsanna Balogh-Brunstad, and Gaël Le Roux, pp. 59–79. Geophysical Monograph. Washington, DC: American Geophysical Union, 2020.

Chapter 7

Kim, Bumsoo, and Yi Ge Zhang. "Methane Hydrate Dissociation across the Oligocene–Miocene Boundary." *Nature Geoscience* 15 (2022): 203–09. doi: 10.1038/s41561-022-00895-5.

Peng, Shushi, Xin Lin, Rona L. Thompson, Yi Xi, Gang Liu, Didier Hauglustaine, Xin Lan, et al. "Wetland Emission and Atmospheric Sink Changes Explain Methane Growth in 2020." *Nature* 612 (2022): 477–82. doi: 10.1038/s41586-022-05447-w.

Ruppel, Carolyn D., and John D. Kessler. "The Interaction of Climate Change and Methane Hydrates." *Reviews of Geophysics* 55 (2017): 126–68. https://doi.org/10.1002/2016RG000534.

Saunois, M., A. R. Stavert, B. Poulter, P. Bousquet, J. G. Canadell, R. B. Jackson, P. A. Raymond, et al. "The Global Methane Budget 2000–2017." *Earth System Science Data* 12 (2020): 1561–623. doi: 10.5194/essd-12-1561-2020.

Stavert, Ann R., Marielle Saunois, Josep G. Canadell, Benjamin Poulter, Robert B. Jackson, Pierre Regnier, Ronny Lauerwald, et al. "Regional Trends and Drivers of the Global Methane Budget." *Global Change Biology* 28 (2022): 182–200. https://doi.org/10.1111/gcb.15901.

Turetsky, Merritt R., Benjamin W. Abbott, Miriam C. Jones, Katey Walter Anthony, David Olefeldt, Edward A. G. Schuur, Guido Grosse, et al. "Carbon Release through Abrupt Permafrost Thaw." *Nature Geoscience* 13 (2020): 138–43. doi: 10.1038/s41561-019-0526-0.

Chapter 8

Hutchins, David A., and Douglas G. Capone. "The Marine Nitrogen Cycle: New Developments and Global Change." *Nature Reviews Microbiology* 20 (2022): 401–14. doi: 10.1038/s41579-022-00687-z.

Ji, Qixing, Erik Buitenhuis, Parvadha Suntharalingam, Jorge L. Sarmiento, and Bess B. Ward. "Global Nitrous Oxide Production Determined by Oxygen Sensitivity of Nitrification and Denitrification." *Global Biogeochemical Cycles* 32 (2018): 1790–802. doi: 10.1029/2018gb005887.

Miner, Kimberley R., Merritt R. Turetsky, Edward Malina, Annett Bartsch, Johanna Tamminen, A. David McGuire, Andreas Fix, et al. "Permafrost Carbon Emissions in a Changing Arctic." *Nature Reviews Earth & Environment* 3 (2022): 55–67. doi: 10.1038/s43017-021-00230-3.

Stein, Lisa Y. "The Long-Term Relationship between Microbial Metabolism and Greenhouse Gases." *Trends in Microbiology* 28 (2020): 500–11. https://doi.org/10.1016/j.tim.2020.01.006.

Tian, Hanqin, Rongting Xu, Josep G. Canadell, Rona L. Thompson, Wilfried Winiwarter, Parvadha Suntharalingam, Eric A. Davidson, et al. "A Comprehensive Quantification of Global Nitrous Oxide Sources and Sinks." *Nature* 586 (2020): 248–56. doi: 10.1038/s41586-020-2780-0.

Voigt, Carolina, Maija E. Marushchak, Benjamin W. Abbott, Christina Biasi, Bo Elberling, Steven D. Siciliano, Oliver Sonnentag, et al. "Nitrous Oxide Emissions from Permafrost-Affected Soils." *Nature Reviews Earth & Environment* 1 (2020): 420–34. doi: 10.1038/s43017-020-0063-9.

Chapter 9

Bai, Yongfei, and M. Francesca Cotrufo. "Grassland Soil Carbon Sequestration: Current Understanding, Challenges, and Solutions." *Science* 377 (2022): 603–08. doi: 10.1126/science.abo2380.

Cavelius, Philipp, Selina Engelhart-Straub, Norbert Mehlmer, Johannes Lercher, Dania Awad, and Thomas Brück. "The Potential of Biofuels from First to Fourth Generation." *PLOS Biology* 21 (2023): e3002063. 10.1371/journal.pbio.3002063.

Harrison, Brendan P., Si Gao, Melinda Gonzales, Touyee Thao, Elena Bischak, Teamrat Afewerki Ghezzehei, Asmeret Asefaw Berhe, et al. "Dairy Manure Co-Composting with Wood Biochar Plays a Critical Role in Meeting Global Methane Goals." *Environmental Science & Technology* 56 (2022): 10987–96. doi: 10.1021/acs.est.2c03467.

Klimasmith, Isaac M., and Angela D. Kent. "Micromanaging the Nitrogen Cycle in Agroecosystems." *Trends in Microbiology* 30 (2022): 1045–55. https://doi.org/10.1016/j.tim.2022.04.006.

Lark, Tyler J., Nathan P. Hendricks, Aaron Smith, Nicholas Pates, Seth A. Spawn-Lee, Matthew Bougie, Eric G. Booth, et al. "Environmental Outcomes of the US Renewable Fuel Standard." *Proceedings of the National Academy of Sciences* 119 (2022): e2101084119. doi: 10.1073/pnas.2101084119.

Logan, Bruce E., Ruggero Rossi, Ala'a Ragab, and Pascal E. Saikaly. "Electroactive Microorganisms in Bioelectrochemical Systems." *Nature Reviews Microbiology* 17 (2019): 307–19. doi: 10.1038/s41579-019-0173-x.

National Academies of Sciences, Engineering, and Medicine. *A Research Strategy for Ocean-Based Carbon Dioxide Removal and Sequestration.* Washington, DC: The National Academies Press, 2022.

Index

For the benefit of digital users, indexed terms that span two pages (e.g., 52–53) may, on occasion, appear on only one of those pages.

Figures and boxes are indicated by *f* and *b* following the page number

4Rs, 147–48, 164
acetate, 114, 152–54, 163–64
aerosols, 73–74, 76, 83, 85, 169–70
agriculture, 36, 139–40, 171–72
  methane source, 121, 122, 132, 162–63
  nitrous oxide source, 132–33, 135, 147–48, 162, 164
  See also biofertilizer; fertilizer; manure
albedo, 48, 76, 83–84, 85, 89
algae, 6, 79–80
  biofuel synthesis, 157–58
  blooms of, 48–49, 67, 138
  carbon dioxide consumption, 3, 24*b*, 62–63
  cloud formation, 75–76, 83, 84–85, 86
  mineral formation, 89–90, 93–94, 97
  primary production, 27
  symbiotic, 93–94, 95, 105–6
  types of, 22–23, 89, 93–94, 95
  See also macroalgae; microalgae; phytoplankton
ammonia, 80, 132–33, 134–35, 148, 149, 164
  nitrogen cycle, 139, 140*f*
  See also ammonia oxidation; nitrification; nitrous oxide
ammonia oxidation, 142*b*, 143–46
ammonium, 93–94, 134–35, 139, 143, 145, 164–66
anaerobes, 114, 153*f*
anaerobic digesters, 152–55
animals, 5–6, 119–21, 141
  carbon cycle, 7–8, 8*f*, 15–16, 29
anoxia, 33–34, 118–19, 129, 138, 147
Antarctica, 1–2, 3, 45–46, 64–65, 84–85
  greenhouse gas source, 124–25, 148–49

ice cores, 40, 63–64, 108–10, 131, 136*f*
anthropogenic
  carbon dioxide, 20–21, 22*f*, 46–47, 173
  methane, 109–10, 111–12, 112*f*, 116–17
  nitrous oxide, 132–33, 133*f*
aragonite, 96, 100
archaea, 114, 134–35, 139, 141, 172
  anaerobic food chain, 152–54, 153*f*, 163–64
  biomethane, 152–54
  definition of, 3–4, 5*f*
  evolution, 4, 79–80, 98
  methane degradation, 119, 129
  methane production, 9, 101, 111, 113–14, 117, 163–64
  nitrous oxide production, 143, 144, 165
  See also methanogenesis; methanogens; methanotrophy
Arctic, 44–46, 98–99, 125
  methane source, 122, 126–29, 127*f*
  nitrous oxide source, 148–49, 150
Arctic Ocean, 84, 119–21, 124–28, 149–50
Arrhenius, Svante, 69, 72–73
atmospheric deposition, 132–33, 133*f*
autotrophy, 6, 16, 26. *See also* primary production

bacteria, 146, 155–57, 171
  abundance of, 14, 37
  biological pump, 50*f*, 53–54, 56
  carbon dioxide release by, 5–6, 7–8, 8*f*
  definition of, 3–4, 5*f*
  dissolved organic carbon use, 54–55, 55*f*, 56–58, 60–61, 69–70
  evolution, 79–80, 98, 99–100, 103
  methane degradation, 117, 118–21, 147, 164

bacteria (*cont.*)
 mineral dissolution, 105–7
 mineral formation, 101–2
 nitrous oxide production, 140*f*, 141,
  143, 144–46, 165
 organic carbon degradation, 31–33, 43–
  44, 45–46, 138, 152
 organic matter production, 37–39, 59,
  60, 173
 respiration, 15–16, 43, 69
 sulfur cycle, 75, 76–77, 114, 119
 symbiotic, 29, 79–80, 119–22, 163–
  64, 176
 See also cyanobacteria; *Prochlorococcus*
ballast, in the biological pump, 53, 89–
 90, 97
bicarbonate, 17, 49, 90, 96, 100–1
biochar, 163–64, 171–72, 173, 175–76
biofertilizer, 165–66, 167
biofuels, 151, 158
 bioethanol, 159–62
 biogas, 152, 154–55, 163
 biomethane, 152–55, 163, 175–76
biogas digesters, 152–55
biological pump, 50–51
 aggregates, 52–53, 61–62, 70
 carbonate pump, 51, 90, 97
 climate change, 62–68, 70
 dissolved organic carbon, 53, 56–59
 marine snow, 50*f*, 51–53, 61–62, 70
 microbial loop, 54–55, 55*f*, 69–71
 zooplankton, 50*f*, 53–54, 58

calcification, 89, 93–94
 ocean acidification, 96–97
 organisms involved in, 92, 98, 99–101
calcite, 100
calcium carbonate
 carbonate pump, 51, 90, 97
 formation of, 89, 92, 93–94, 98–99, 100–1
 microbial mats, 99 –102
 ocean acidification, 96
 weathering of, 104–5, 104*f*
 See also carbonate rocks; chalk;
  dolomite; limestone
carbonate pump, 51, 90, 97
carbonate rocks, 13–14, 89–90, 97–100,
 103, 104–6

carbon cycle
 changes in, 27, 39–40, 45–46, 128–
  29, 151
 components of, 7–8, 8*f*, 12–14, 13*f*, 22*f*
 fossil fuels, 12, 13*f*, 22*f*, 28
 oceanic, 21, 54–55, 67, 68, 97
 slow cycle, 90–91, 92, 103, 104–5, 107
 terrestrial, 29, 35, 39, 43–44, 47
carbon dioxide, 8–9
 atmospheric, 2*f*, 13*f*, 22*f*, 27, 34, 50
 chemistry in water, 49, 96, 104–5
 consumption of, 21, 27
 fertilization by, 40–42, 43, 62–63
 from fossil fuels, 20–21, 22*f*
 production of, 5–6, 20–21
 and oxygen, 5–6, 7, 8*f*, 20, 34
 sequestration of, 169, 170–71,
  170*f*, 174
Carboniferous, 31–34, 106–7
carbonic acid, 73, 104–5
carbon sequestration
 geoengineering, 169, 170–71, 174
 oceanic, 50, 65–67, 70–71, 170*f*, 174
 terrestrial, 170*f*, 173, 175–76
Carson, Rachel, 51–53
cattle, 122, 133
 methane from, 111–12, 121, 162–64
corn, 136–37, 138, 145, 147–48, 159–
 61, 165–67
chalk, 87–88, 89–90, 92
chemolithotrophy, 141–43, 142*b*, 146
China, 136–37, 154–55, 159, 162–63
 carbon studies in, 15, 38, 41
 coal use, 111, 135
chlorofluorocarbons (CFCs), 10–11, 74–
 75, 131
chlorophyll, 6, 25, 63, 64–65, 81–82,
 85, 174
clathrates, 123–24. *See also* hydrates
CLAW hypothesis, 77, 81–82, 83–85. *See
 also* dimethyl sulfide (DMS); Gaia
climate sensitivity, 72, 73–74
cloud condensation nuclei (CCN), 73–74,
 75–76, 77, 82–83, 85
clouds, 48, 169–70, 170*f*
 albedo, 76, 83–84, 85
 formation of, 73–74, 75–76, 78, 81, 83
 microbes, 74, 76–77, 82*f*, 85

coal, 31–33, 167–68
  formation of, 31–34
  methane from, 109–10, 111
coccolithophores, 84–85, 90–91, 98, 103
  biological pump, 53, 89–90
  calcium carbonate, 51, 88, 89–90, 92,
    97, 100
  ocean acidification, 96, 97–25
coccoliths, 88, 89, 90–91, 97
COVID-19, 20–21, 108–9, 110
corals, 48–49, 89, 92–98, 102, 103
Cretaceous, 33, 87–88
cyanobacteria, 79–80
  definition of, 5f, 6
  harmful blooms of, 67, 137–38
  microbial mats, 99–102
  nutrients, 22–23, 67, 84–85, 137
  primary production, 21, 22–23, 27
  types of, 22–24, 137–38

Davy, Humphry, 130–31, 155
dead zones, 68, 138
deep time, 90–92, 105
deep water formation, 49–50, 58
denitrification, 149, 150
  definition of, 139–40, 140f
  nitrous oxide degradation, 146
  nitrous oxide production, 139, 140–41,
    145–46, 148, 149, 150, 164, 165
Devonian, 31–33, 106–7
diatoms, 22–23, 53, 64–65, 67, 84–
  85, 89–90
diffusion, 49–50, 62, 101, 113–15, 117–18
dimethyl sulfide (DMS), 9, 169–70, 170f
  cloud condensation nuclei, 76–77, 77f
  Gaia, 81–85, 82f
dimethylsulfoniopropionate
  (DMSP), 76–78, 81–82, 84–85
dinoflagellates, 93–94, 95, 103
diseases, 4, 11, 94, 108
dissolved inorganic carbon (DIC), 13f,
  22f, 49. See also bicarbonate;
  carbonic acid
dissolved organic carbon (DOC), 13f, 54, 83
  labile, 54–55, 56–58, 69–70
  refractory, 56, 58–62, 69–70
  semi-labile, 56–58, 57f
dolomite, 92–94, 98–99, 101

El Niño, 95, 116–17
Emiliania huxleyi, 89, 97
Eocene, 63–64, 123–24
ethanol, 158–61. See also biofuels
eukaryotes, 22–23, 51, 62–63, 79–
  80, 84–85
  definition of, 3–5, 5f
  physiology of, 134–35, 139–40,
    141, 146
Europe, 136–37, 162–63
  biofuels, 154–55, 157–58, 159
  carbon studies in, 15, 38, 40, 41, 46–47
European Union, 108, 154–55
export production, 65–68, 70. See also
    biological pump

fermentation, 152–54, 158–61
fertilizer, nitrogen
  history of, 133–37
  microbial sources of, 165–66, 167
  nitrous oxide from, 133–34, 136f, 145,
    147–48, 160
  pollution, 136–38
food chains, 26, 67–68, 69–70, 137
  biological pump, 54–56
foraminifera, 89, 92, 97–98, 103
fossil fuels, 2–3, 135, 156–57, 158,
  167, 175–76
  carbon budget, 14, 135–36
  carbon dioxide from, 20–21, 22f, 28, 96
  carbon in, 13–14, 13f, 103
  climate change, 1, 12, 28, 72, 151
  methane from, 109–10, 112f, 113, 121
  nitrous oxide from, 132–33, 133f
  See also coal; methane; natural gas
fungi, 3–4, 5f, 41, 175–76
  abundance of, 37, 38, 54–55
  lichens, 105–7
  lignin degradation, 31, 33–34
  nitrogen cycle, 139–41, 146
  respiration, 5–6, 5f, 15–16, 43
  soil organic matter formation, 38,
    39f, 171
  symbiotic, 30–31, 35, 41, 121, 170–71
  weathering, 105–7

Gaia, 9, 78–83, 85–86
geoengineering, 169–71, 173–74, 175

Germany, 92–93, 119, 125, 134, 159
global warming, 8–10, 46–47, 108,
     169, 171–72
  coral reefs, 95, 96
  methane, 116–17, 123–26, 128–29
  nitrous oxide, 148–50
  ocean, 62, 65–66, 67, 68–69, 70, 76
  Paleocene-Eocene Thermal Maximum,
     63–64, 123–24
  permafrost, 44, 47, 128–29
  respiration, 42–44, 69–70
Great Barrier Reef, 93–94, 102
Great Oxidation Event, 6–7
green energy, 151. *See also* biofuels
greenhouse gases, 3, 14–15, 74, 162
  Gaia, 77, 80, 86
  types of, 1, 8–11

Haber-Bosch process, 134–37,
     165–67
halogenated gases, 10–11. *See also*
     chlorofluorocarbons (CFCs)
harmful algal blooms, 48–49, 67, 138
heterotrophy, 8*f*, 105–6
  definition of, 5–6, 142
  nitrogen cycle, 140–41, 145–46
  oceanic, 26, 54–55
  See also respiration
high nutrient-low chlorophyll (HNLC),
     64–65, 174
humification, 35–36, 37, 59–60
humus, 35–37, 59–60
hydrates, 119, 122–29
hydrogen gas, 114, 134–35, 152–54
hydrogen sulfide, 7, 75, 101, 152–54
hydrological cycle, 9–10, 105, 116, 122–
     23, 172
hydrothermal vents, 59, 60–61, 103, 119–
     21, 139

ice ages, 65, 72–73, 85–86, 131
  carbon dioxide level, 27, 34
  methane hydrates, 123–24, 126–27
  Southern Ocean, 63–64, 65, 174
  weathering, 105, 106–7
Intergovernmental Panel on Climate
     Change (IPCC), 39–40, 73,
     108, 171–72

carbon cycle, 13–14, 13*f*, 22*f*, 122–
     23, 126–27
  predictions by, 45–46, 47, 150, 167
iron, 7, 64–65, 84–85, 105–6, 169–
     70, 174–75

Keeling Curve, 1–3, 2*f*, 12
Keeling, Charles David, 1–3, 11

laughing gas, 9, 130–31
Leeuwenhoek, Antonie van, 3–4, 91–92
legumes, 29, 134–35, 165–66
lichens, 105–7
lignin, 31–34, 154, 160–61
limestone, 92–94, 98–100, 101, 111
Lovelock, James, 74–76, 78–81, 86

macroalgae, 157–58, 163–64, 173
maize, 136–37, 159. *See also* corn
manure
  methane, 111–12, 151–54, 163
  nitrous oxide, 133, 145–46, 164
Margulis, Lynn, 78–81
marine heatwaves, 48–49, 95
marine snow, 52–53, 61–62, 70
Martin, John, 64–65, 174
methane, 7, 101, 135, 139, 151
  atmospheric level, 8–9, 109–10, 109*f*
  biomethane, 152–55, 157–58,
     163, 175–76
  degradation of, 110, 117–22, 125–26, 147
  Gaia, 78–79, 79*f*, 80
  greenhouse effect, 8–9, 108
  hydrates, 122–29
  permafrost, 45–46, 126–29
  reducing emissions of, 162–64, 172
  sources of, 4, 110–17, 162
  symbioses, 119–21, 122
methanogenesis, 114, 116, 119, 125,
     129, 152–54
methanogens, 4, 110–12, 129
  habitats of, 114, 116, 118*f*, 126–
     27, 152–54
  symbioses, 121–22, 163–64
methanotrophy, 117–18, 124–25, 164
  hydrates, 124–26, 128
  symbiosis, 119–21
  wetlands, 117–18, 118*f*

microalgae, 16, 51, 62–63, 157–58, 167–
    68. *See also* algae; phytoplankton
microbe, definition of, 3–4. *See also*
    archaea; bacteria; fungi; microalgae;
    protists; protozoa; viruses
microbial carbon pump, 49, 58–59, 61–62,
    63–64, 69–70
microbial fuel cells, 155–57
microbial loop, 54–56, 69–71
microbial mats, 99–102, 103, 139
microbiomes, 121–22, 144, 163–64, 166–
    67, 176
Milankovitch cycles, 34, 65
mountain formation, 92–94
mycorrhiza, 30–31, 34–35, 41, 170–71

natural gas, 135, 154–55
    methane source, 108, 109–10, 115, 119–
    21, 122–23
necromass, 38–39, 173
negative emissions, 167, 171–72
negative feedback, 81, 82–83
nitrate, 133–35, 139–40, 144–45, 146
    nitrous oxide source, 139, 140f, 143,
        145, 148, 164
    nutrient, 18, 64–65, 66f, 137, 145
    pollution, 137–38
nitric oxide, 144–45
nitrification, 142, 143–44, 150, 165
    comammox, 143, 144–45, 146
    definition of, 139, 140f, 141
    global warming effect on, 148, 149
    nitrous oxide production, 140f, 143,
        144, 145
nitrite, 143, 144–46
nitrogen cycle, 7–8, 131–32, 139, 140f, 150.
    See also denitrification; nitrogen
    fixation; nitrification
nitrogen fixation, 29, 35, 134–35, 165–67
nitrogen gas, 36, 134–35, 139–41,
    146, 165–66
    atmospheric, 78–79, 79f, 80, 86
nitrogen pollution, 136–38, 160
nitrogen use efficiency, 136–37, 147–48
nitrous oxide, 9, 45–46
    atmospheric, 8–9, 80, 131, 136f
    degradation of, 146–47
    global warming, 148

greenhouse effect by, 8–9, 131
    history of, 130–31
    permafrost, 148–50
    production of, 139–46
    sources of, 132–34, 133f, 147–48

ocean, 7, 48–49, 74, 143
    carbon in, 13–14, 13f, 35
    carbon uptake by, 12, 22f, 49, 62, 70–71
    geoengineering, 173–75
    methane, 113–14, 119–21, 124–26
    nitrous oxide, 132–33, 133f, 145, 147, 148
    primary production in, 21, 25–26, 27,
        63–65, 67–68
    respiration in, 16, 20, 26–27, 69–70
    See also Arctic Ocean; Southern Ocean
ocean acidification, 11, 96, 97, 148, 169–70
ocean deoxygenation, 68, 114
organic matter, 35–36, 56, 69, 83, 138
    biological pump, 51, 53–54, 61–62
    carbon cycle, 7–8, 8f
    degradation of, 114, 139–40, 141
    methane, 152–54
    nitrous oxide, 139–40, 145–46, 148, 164
    See also dissolved organic carbon
        (DOC); refractory organic matter;
        soil organic matter (SOM)
oxygen, 45–46, 60–61, 94, 148, 152–54
    atmospheric, 6–7, 34, 78–79, 79f
    carbon cycle, 5–6, 8f
    carbon rate estimation, 17, 20, 26, 63–64
    Gaia, 78–79, 79f
    lignin degradation, 31, 33–34
    methane, 114–15, 117–18, 163
    nitrous oxide degradation, 146, 147
    nitrous oxide production, 139–41,
        143, 144–45
    organic matter degradation, 129, 145–
        46, 172
    sulfur cycle, 75, 101, 119, 139
oxygen minimum zones (OMZs), 68,
    147, 148

peat, 13–14, 46–47, 117, 119–21, 172–73
permafrost
    carbon in, 13–14, 13f, 39, 44
    global warming of, 44–46, 47
    methane, 126–29, 127f

Phaeocystis, 77, 84–85
photosynthesis, 85–86, 100–1, 106
    carbon dioxide, 2–3, 6–7, 8*f*, 14–15,
        21, 24, 49
    enzyme key in, 40–41, 62–63
    organic carbon, 6, 54, 93–94, 99–100,
        99*b*, 171
    organisms, 6, 79–80
    oxygen from, 6–7, 17, 18, 19–20, 68–69
phytoplankton, 27, 85–86
    biological pump, 50*f*, 51, 53, 61–62
    CLAW hypothesis, 76–78, 81–82, 83, 85
    climate change effects on, 62–64, 67–
        68, 97
    dimethyl sulfide, 75–76, 77, 81–
        82, 84–85
    dissolved organic carbon, 54–55,
        55*f*, 59–60
    geoengineering, 170*f*, 174–75
    primary production by, 24, 25–27
    types of, 22–24, 89
plants, 23–24, 106–7, 157–58
    agricultural, 136–37, 147–48,
        164, 165–67
    bioethanol, 159–61
    carbon dioxide consumption, 2–3, 7–8,
        8*f*, 22, 22*f*, 29
    carbon dioxide fertilization, 40–41
    carbon in, 5, 13*f*, 14
    detritus from, 31–34, 45–47
    geoengineering, 170–71, 173
    microbial symbioses, 29–31, 30*f*
    physiology of, 6, 79–80
    refractory organic carbon, 31–33, 35–
        36, 37, 39*f*
    respiration, 15–16, 22*f*, 42–43
    soil organic matter, 37, 39*f*, 43
    wetlands, 114–15, 117–18, 118*f*, 172
positive feedback, 42, 81, 116–17, 148
Precambrian, 98–100, 102–3, 141
precipitation, 132–33, 148, 149–50
    hydrological cycle, 9–10, 116–17
primary production, 14–15, 20, 89–90, 174
    carbon cycle, 21, 28
    climate change, 40–41, 45–46, 47,
        63, 65–66
    gross, 18, 21–22, 24, 25–26, 27, 40–41
    net, 17, 19–20, 27

net community, 18, 21–22
    oceanic, 22–24, 25–27, 54–56
    organisms, 21, 27, 89
    terrestrial, 21–22, 35, 37, 40, 41,
        47, 63–64
Prochlorococcus, 23–24, 54–55
prokaryotes, 3–4, 9, 14, 22–23, 79–80, 98
protists, 3–5, 53–55, 60, 121, 150
protozoa, 3–4, 30, 38–39, 68, 163–64.
        See also protists

reforestation, 170–71, 173, 175–76
refractory organic matter, 83, 103,
        173, 175–76
    age of, 35–36, 56
    climate change, 43–44, 69–70
    humus, 35–37, 59–60
    microbial carbon pump, 49, 58–
        62, 69–70
    oceans, 56–62, 57*f*, 69–70
    soils, 35–36, 37
    See also dissolved organic carbon
        (DOC); soil organic matter (SOM)
renewable energy, 161–62, 161*f*, 167.
        See also biofuels; biogas digesters
respiration, 27, 28, 105–7, 142*b*
    Arctic, 45–47
    carbon cycle, 8*f*, 14–15, 21, 22*f*
    global warming, 41–44, 69–70
    oceanic, 16, 20, 24, 26, 55–56
    organisms, 5–6, 15–16, 43–44
    soils, 15–16, 41–44, 69
rice, 111–12, 117, 122–23, 129, 164, 165–
        66, 171–72
rumen, 122, 163–64
ruminants, 111–12, 116, 117, 121, 129,
        133, 162–64

Saccharomyces, 158–61
saprotrophs, 31–33, 34, 43–44
sedimentary rocks, 13–14, 89–90, 98,
        101, 103. *See also* carbonate rocks;
        chalk; dolomite; limestone
Siberia, 44–45, 126, 128
silica, 89–90
silicate, 7–8, 64–65, 104–7
Snowball Earth, 85–86, 123–24
soil, 7, 22*f*, 46–47, 114–15, 165, 172

carbon content, 13–14, 13*f*, 39, 44, 103
geoengineering, 170*f*, 171–72, 173
moisture, 46–47, 148, 149–50
nitrous oxide, 132, 133, 133*f*, 140–41, 145, 147
organisms in, 15–16, 30–31, 143, 165–66
See also peat; permafrost
soil organic matter (SOM), 136–37
carbon in, 13–14, 13*f*, 35
carbon sequestration, 106–7, 170*f*, 171, 173, 175–76
formation of, 35–39, 170*f*, 171–72
solubility pump, 49–50, 51, 58, 59, 62–63
Southern Ocean, 63–65, 77, 84–85, 169–70, 174
stratification, water column, 65–67, 85, 148
stromatolites, 101–3
sulfate reduction, 101, 114, 122, 152–54, 163
sulfur, 99–100, 115, 141–43, 169–70
cycle of, 7–8, 75
hydrothermal vents, 119–21, 139
pollutants, 110, 167–68
sulfuric acid, 75–76, 106, 163
weathering, 105–6
See also dimethyl sulfide (DMS); dimethylsulfoniopropionate (DMSP)
symbioses, 79–80, 105–7
corals, 93–95
methane oxidizers, 119–21
methanogens, 121, 122, 163–64

nitrogen fixers, 29, 134–35, 165–66
plants, 29, 30–31, 30*f*

termites, 121
tropics, 98–99, 105, 115, 116, 122, 123–24
tundra, 45–46, 117, 126, 149–50

United States, 108, 138, 162–63
carbon cycle studies in, 40, 41
corn ethanol, 159–60
fertilizer use, 134, 136–37, 136*f*
renewable energy, 154–55, 157–58, 161–62, 161*f*
urea, 139, 145–46, 164–65

viruses, 38–39, 55–56

Waksman, Selman, 36–37, 38–39, 141–43
water vapor, 9–10, 72, 73–74
weathering, 7–8, 34, 98, 104–7
wetlands, 172–73
methane degradation, 117–18, 118*f*
methane emissions, 113, 114–17, 122–23, 129
Winogradsky, Sergei, 141–43
Woese, Carl, 4, 9

yeast, 3–4, 141–43, 152–54, 155, 158–61, 163. See also *Saccharomyces*

zooplankton, 16, 20, 55*f*, 93–94
biological pump, 50*f*, 53–54
predation by, 76–77, 84–85, 88